**MODULAR SERIES
ON SOLID STATE DEVICES**

W9-BLB-137

# VOLUME III

# The Bipolar Junction Transistor

MODULAR SERIES
ON SOLID STATE DEVICES

Gerold W. Neudeck and Robert F. Pierret, Editors

# VOLUME III
# The Bipolar Junction Transistor

## GEROLD W. NEUDECK

Purdue University

▲
▼▼

ADDISON-WESLEY PUBLISHING COMPANY

READING, MASSACHUSETTS
MENLO PARK, CALIFORNIA
LONDON · AMSTERDAM
DON MILLS, ONTARIO
SYDNEY

This book is in the
**Addison-Wesley Modular Series on Solid State Devices**

**Library of Congress Cataloging in Publication Data**

Neudeck, Gerold W.
  The bipolar junction transistor.

  (Modular series on solid state devices ; v. 3)
  Bibliography: p.
  Includes index.
  1. Bipolar transistors.   2. Junction transistors.   I. Title.
II. Series: Pierret, Robert F. Modular series on solid state
devices ; v. 3.
TK 7871.96.B55N48      621.3815'282      81-14977
ISBN 0-201-05322-5                  AACR2

ISBN 0-201-05322-5
ABCDEFGHIJ-AL-898765432

# Foreword

Solid state devices have attained a level of sophistication and economic importance far beyond the highest expectations of their inventors. The bipolar and field-effect transistors have virtually made possible the computer industry which in turn has created a completely new consumer market. By continually offering better performing devices at lower cost per unit, the electronics industry has penetrated markets never before addressed. One fundamental reason for such phenomenal growth is the enhanced understanding of basic solid state device physics by the modern electronics designer. Future trends in electronic systems indicate that the digital system, circuit, and IC layout design functions are being merged into one. To meet such present and future needs we have written this series of books aimed at a qualitative and quantitative understanding of the most important solid state devices.

Volumes I through IV are written for a junior, senior, or possibly first-year graduate student who has had a reasonably good background in electric field theory. With some deletions these volumes have been used in a one semester, three credit-hour, junior-senior level course in electrical engineering at Purdue University. Following this course are two integrated-circuit-design and two IC laboratory courses. Each volume is written to be covered in 12 to 15 fifty-minute lectures.

The individual volumes make the series useful for adoption in standard and non-standard format courses, such as minicourses, television, short courses, and adult continuing education. Each volume is relatively independent of the others, with certain necessary formulas repeated and referenced between volumes. This flexibility enables one to use the series continuously or in selected parts, either as a complete course or as an introduction to other subjects. We also hope that students, practicing engineers, and scientists will find the volumes useful for individual instruction, whether it is for reference, review, or home study.

A number of the standard texts on devices have been written like encyclopedias, packed with information, with little thought to how the student learns or reasons. Texts that are encyclopedic in nature are difficult for students to read and are often barriers to their understanding. By breaking the material into smaller units of information and by

v

writing for students (rather than for our colleagues) we hope to enhance their understanding of solid state devices. A secondary pedagogical strategy is to strike a healthy balance between the device physics and practical device information.

The problems at the end of each chapter are important to understanding the concepts presented. Many problems are extensions of the theory or are designed to reinforce particularly important topics. Some numerical problems are included to give the reader an intuitive feel for the size of typical parameters. Then, when approximations are stated or assumed, the student will have confidence that certain quantities are indeed orders of magnitude smaller than others. The problems have a range of difficulty, from very simple to quite challenging. We have also included discussion questions so that the reader is forced into qualitative as well as quantitative analyses of device physics.

Problems, along with answers, at the end of the first three volumes represent typical test questions and are meant to be used as review and self-testing. Many of these are discussion, sketch, or "explain why" types of questions where the student is expected to relate concepts and synthesize ideas.

We feel that these volumes present the basic device physics necessary for understanding many of the important solid state devices in present use. In addition, the basic device concepts will assist the reader in learning about the many exotic structures presently in research laboratories that will likely become commonplace in the future.

*West Lafayette, Indiana*

G. W. Neudeck
R. F. Pierret

# Contents

# 3 Deviations from the Ideal Transistor

# 4 Small Signal Models

# 5 Switching Transients

# Suggested Readings

# Volume Review Problems and Answers

# Appendix

# Index

# Introduction

This volume concentrates on the bipolar junction transistor (BJT), presenting both a qualitative and quantitative description of the device. The two basic forms of the bipolar transistor are the *n-p-n* and *p-n-p*, named from the three layers of semiconductor used to construct the device. Both forms are in general use as individual discrete devices and in integrated circuits. Because of its generally higher gain and faster switching, the *n-p-n* version is preferred in many circuit designs.

To make the bipolar presentation efficient, many of the concepts and derivations developed in Volume II for the junction diode are applied directly to the junction transistor. However, the bipolar transistor differs from the diode in that the transistor is capable of current gain, voltage gain, and power gain. It is an active device whereas the diode, like a resistor, is a passive device. We therefore emphasize the ability of the transistor to achieve gain.

The *n-p-n* and *p-n-p* bipolar transistors are "*complementary.*" This means that the devices have an interchange of material types (*n* to *p* and *p* to *n*) and have opposite polarities for all voltages and currents. We will use this idea of complements, rather than repeat the physical descriptions, currents, and voltages for both the *n-p-n* and *p-n-p* device. The following chapters use the *p-n-p* device in developing the concepts and equations describing the bipolar transistor, because the directions of the currents and carrier fluxes are simpler to describe. Since the *n-p-n* is the complement of the *p-n-p* one needs only to change all current and voltage polarities to describe the *n-p-n*.

Chapter 1 establishes the reference directions for the currents and voltages as well as a qualitative description of bipolar transistor action and fabrication. Chapter 2 outlines a "game plan" for the solution of carrier concentrations in the base and the equations necessary to obtain the terminal currents as a function of the terminal voltages. An ideal bipolar transistor is defined and the terminal currents are derived. Chapter 2 also discusses the regions of operation of the ideal *p-n-p* and presents the common emitter input and output $V-I$ characteristics. The dc values for alpha, beta, emitter injection efficiency, $I_{CB0}$, and $I_{CE0}$ are derived and discussed qualitatively. The Ebers–Moll equations for the ideal device are derived and are noted to be valid for all four regions of operation. These

1

equations embody the nonlinear dc model for bipolar transistor devices, and serve a role similar to the ideal diode equation for the diode.

The deviations of a real device from the ideal transistor are the subject of Chapter 3. Base width changes, avalanche breakdown, reach-through, generation–recombination, and graded base effects are discussed. Chapter 4 derives the small signal, low-frequency, hybrid-pi model for the ideal transistor. Nonideal effects are then included to complete the low-frequency model. The depletion and diffusion capacitances for the device are summarized and added to obtain a high-frequency signal model. Finally, Chapter 5 presents the charge control model and derives the storage time for a saturated transistor that is switched to cutoff with a base current step.

# 1 / Introduction to Bipolar Junction Transistors

## 1.1 TERMINOLOGY AND SYMBOLS

The junction transistor is, by definition, a semiconductor device containing three adjoining, alternately doped regions in which the middle region is very narrow. As shown in Fig. 1.1(a), the external world contact to the narrow central region is known as the *base* . Contacts to the outer regions are labeled the *emitter* and *collector* . The emitter and collector designations arise from the functions performed by these regions in the operation of the device; it would appear from Fig. 1.1(a) that these two regions might be interchangeable. However, in modern day practical devices, the emitter region is usually much more heavily doped than the collector and the terminals cannot be interchanged without changing the device's characteristics.

Figure 1.1(b) illustrates the circuit symbol for the *pnp* junction transistor while simultaneously defining the pertinent current and voltage polarities. Although "+" and "−" signs are shown in this figure to define voltage polarities, they are actually redundant, because the double subscript on the voltage symbol likewise denotes the voltage polarity. The first subscript indicates the assumed reference polarity of "+". For example $V_{EB}$ assumes the "E" to have the "+" sign and "B" the "−" sign. Note that as a consequence

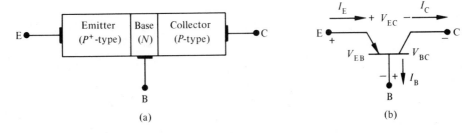

**Fig. 1.1** *pnp* bipolar transistor: (a) semiconductor types; (b) circuit symbol with active region voltage and current polarities.

of Kirchhoff's circuit laws there are only two independent voltages and two independent currents. If two currents or voltages are known, then the third is also known.

Throughout this volume, current reference directions were chosen to coincide with the physical current flow in the active region for each device rather than the IEEE standard notation of all currents for all devices entering each terminal. With the current reference directions as chosen, all currents in the active region are positive quantities; this makes conceptual arguments much easier for the reader. It also provides a more intuitive base for understanding the device physics of carrier flow.

Figure 1.2(a) illustrates the "complement" of the *pnp*, the *npn* transistor. By complement we mean the interchange of *p* for *n* and *n* for *p*. The current and voltage reference directions are displayed in Fig. 1.2(b) for the *npn*. Comparing Fig. 1.1(b) and Fig. 1.2(b), it is clear that to obtain the complement of the *pnp*, the *npn*, all the currents and voltage polarities are reversed. If you understand the *pnp*, then you need only reverse polarities and conduction types to describe the *npn*. This volume will concentrate on the *pnp*, because it follows more directly the format of the equations and polarities developed in Volume II for the junction diode.

The *pnp* transistor of Fig. 1.1(a) can be thought of as two very closely spaced *p-n* junctions. One junction is formed between the emitter and base, the other between the collector and base regions. The *n*-type base region is typically less than 1 micron in width. Because of the close proximity of the two junctions they interact with each other, and thus the transistor is capable of current and voltage gain. A more detailed discussion of their interaction and gain is a topic of Chapter 2.

The bipolar transistor has four *regions of operation*. By regions of operation we mean the voltage polarities on the collector to base junction and the emitter to base junction. For example, the junction diode had two regions of operation, forward and reverse bias, depending on the junction voltage polarity. The most common region of operation for the bipolar transistor is the *active region* defined as having the E−B junction forward biased and the C−B junction reverse biased. For the *pnp* this means the E−B has a "+" to "−" polarity and the C−B has a "−" to "+" polarity. Almost all linear signal amplifiers have their bipolar transistors biased in the active region because they have their largest signal gain in this region.

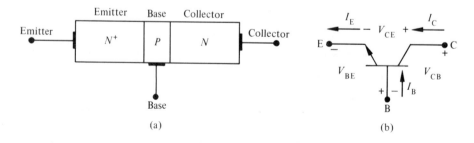

(a)                                              (b)

**Fig. 1.2**   *npn* bipolar transistor.

The *saturation region* is defined as having both the E–B junction and the C–B junction forward biased. For the *pnp* this means a positive $V_{EB}$ and $V_{CB}$ voltage. In logic circuits and transistor switches, this represents the region of operation where $|V_{CE}|$ is small and $|I_C|$ is large; that is, the device acts like an "on" switch. A closed (or "on") switch has little or no voltage drop across it, that is, a logic level of "low."

The *cutoff region* is defined as having both junctions reverse biased. A negative voltage polarity of $V_{EB}$ and $V_{CB}$ is necessary for the *pnp* transistor. Typically this represents the "off" state for the transistor as a switch, or the "high" logic level in digital circuits. When "off," the transistor is similar to an open switch in that $|I_C|$ is nearly zero and $|V_{CE}|$ is large.

The fourth region of operation is the *inverted region*, which is sometimes called the *inverted active region*. For inverted operation the E–B is reverse biased and the C–B is forward biased. One might think of the "inverted active" case as where the collector acts like the emitter and the emitter like a collector; that is, the device is used backwards. The most common use of this region of operation is in digital logic circuits such as TTL logic where signal gain is not an objective.

Figure 1.3(a) illustrates the $V_{EB}$ and $V_{CB}$ voltages for the four *pnp* regions of operation. Note that the "complement" of Fig. 1.3(a) for the *npn* would have $V_{BE}$ and $V_{BC}$ for the axes. Figure 1.3(b) illustrates the regions of operation on the output *V–I* characteristic. Chapter 2 discusses each of these regions in detail.

In circuit applications the transistor typically functions with a common terminal between the input and output, either dc common or signal ground common. Because the transistor has only three leads, there are three possible amplifier types. The designations are *common base, common emitter,* and *common collector*; the names indicate the lead common to both the input and output circuits as illustrated in Fig. 1.4(a), (b), and (c) respectively for the *pnp*. The most often used of the three amplifier types is the common emitter. In describing the amplifier and its *V–I* characteristics the current in the "common"

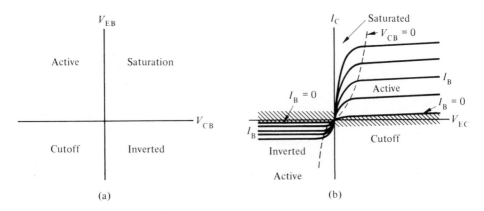

**Fig. 1.3**   *pnp* regions of operation: (a) junction polarities; (b) output characteristic.

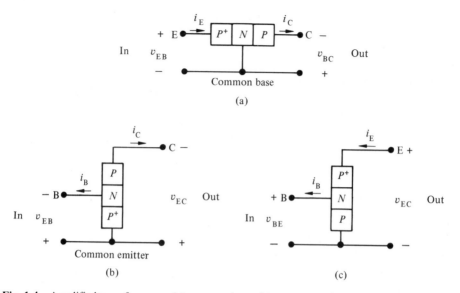

**Fig. 1.4**  Amplified types for a *pnp*: (a) common base; (b) common emitter; (c) common collector.

lead is the variable eliminated. For example the common emitter amplifier has as output variables $v_{EC}$ and $i_C$. Its input variables are $v_{EB}$ and $i_B$. Again note that if two of the voltages (or currents) are known the third is also known from Kirchhoff's Laws. Therefore to describe the common emitter by its output and input $V–I$ characteristics implies that you can also deduce the common base and common collector characteristics.

## 1.2  QUALITATIVE OPERATION OF THE ACTIVE REGION *pnp*

Before one can understand how a transistor works it is necessary to establish a few basic facts. To do this let us consider a *pnp* transistor in thermal equilibrium. Figure 1.5(a) illustrates the energy band diagram for uniformly doped regions where the emitter is doped heavier *p*-type than the collector. The magnitude of the *n*-type base doping is less than the emitter, but greater than the collector. Note that this figure is just an extension of the thermal equilibrium case for the *p-n* junction as applied to a $p^+$-*n* and a *n-p* junction closely spaced. All of the physical arguments presented in Volume II, Chapter 2, are applicable to the E–B and B–C junctions. The built-in potential ($V_{bi}$), charge density, electric fields, etc., as discussed are still valid. Figures 1.5(b) and (c) illustrate the charge density and electric fields in the two depletion regions. At thermal equilibrium there is no net current flow; hence as in the junction diode all the drift and diffusion currents cancel each other out. Said another way, all the drift components are equal and opposite of the diffusion components throughout the device. If ideal junctions are assumed, then all the depletion widths, electric fields, etc., are calculated from the formulas presented for the ideal diode.

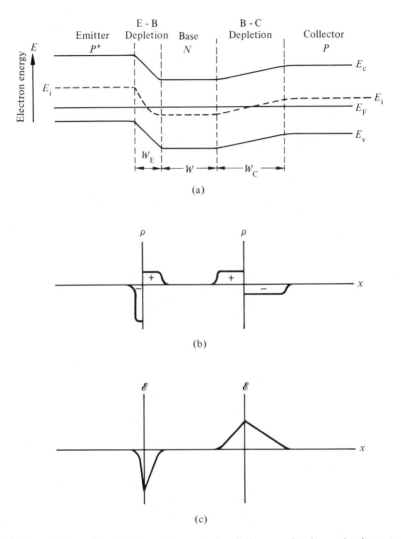

**Fig. 1.5** *pnp* in thermal equilibrium: (a) energy band diagram; (b) charge density; (c) electric field.

The *pnp* device biased in the active region requires the emitter to have a higher potential than the base. Figure 1.6(a) shows the energy band diagram for thermal equilibrium and for active region operation. Note that the barrier for holes entering the *n*-base region from the emitter is lowered. As in a forward biased diode, holes are allowed to be injected from the emitter into the base. Also note that the barrier for electrons in the base is lowered and electrons are injected from the base into the emitter. The combined effect is that the forward biased E–B junction creates a positive emitter current.

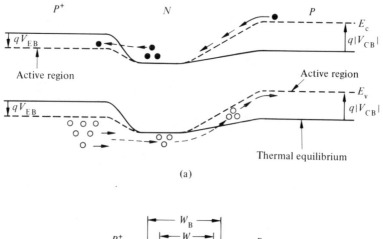

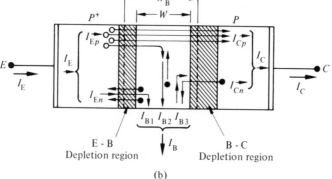

**Fig. 1.6** $p^+np$ in the active region: (a) energy band diagram; (b) current components.

The collector-base junction is reverse biased for active region operation and the right-hand side (the collector) of Fig. 1.6(a) is raised with respect to the $n$-base region. This effectively raises the barrier for electrons in the base that may want to travel to the collector. Similarly the hole barrier from collector to base is increased, reducing any carrier flow. However, note that the holes injected from the emitter into the very narrow ($< 1$ $\mu$m) base easily diffuse through the base and find a potential hill to slide down* into the collector. The reader may now get an inkling of the terminology of emitter (injector of carriers) and collector (as collector of those injected elsewhere). If almost all the injected holes from the emitter traverse the base with $< 1\%$ recombining in the $n$-base region, then the collector hole current is slightly less than the emitter hole current. The electron component of the collector current arises from the thermally generated minority electrons at the edge of the $p$-type collector that fall down the potential hill.

---

*Remember that holes like to "float" and electrons to sink in an electron energy band diagram.

For any "interaction" to take place between the two transistor junctions, they must be spaced such that the base width is much less than the minority (holes in the base) carrier diffusion length. If not the case, then the *pnp* structure is nothing more than two back-to-back diodes; that is, all the holes injected from the emitter recombine in the base and never reach the reverse biased B–C junction.

Looking inside the device, what do we see as far as the electrostatics are concerned? First, around both the emitter–base and collector–base metallurgical junctions there is a depletion region. The E–B space charge region will be smaller than in equilibrium because this junction is forward biased. The C–B junction depletion region will be wider than in equilibrium because this junction is reverse biased. Hence, we can visualize the situation to be something like that shown in Fig. 1.6(b). Note that part of the base, the width $W$, is quasi-neutral; that is, it contains essentially no electric field. It should be emphasized that the width $W$ is *not* equal to the metallurgical base width, $W_B$; specifically, $W$ is the width of the base less the regions of the base that are depleted.

In the preceding discussion we concentrated on the hole current flowing from the emitter to the collector to establish several important facts. Further qualitative information can be gained by examining the base current components in the system. This will be done with the aid of Fig. 1.6(b), which diagrams all the various current components (other than recombination–generation currents in the depletion region, which we ignore for the time being). The three current components shown in this figure have been labeled $I_{B1}$, $I_{B2}$, and $I_{B3}$. Together they make up the total current going out of the base. $I_{B1}$ corresponds to the current arising from electrons being back injected across the forward biased E–B junction from the base into the emitter. $I_{B2}$ corresponds to those electrons that must enter the base to replace electrons used up in recombining with holes injected from the emitter. Only a very small number of injected holes will recombine, with most passing on through the base into the collector. $I_{B3}$ is that part of the collector current due to thermally generated electrons in the collector that are within one diffusion length of the C–B junction edge and that fall down the potential hill from the collector into the base.

Practical transistors are fabricated so that the emitter doping is much greater than the base doping, which in turn is made much greater than the collector doping. What effect does this have on the size of the current in the transistor? Because the emitter doping is much greater than that of the base, the hole current across the E–B junction ($I_{Ep}$) will be much greater than the electron current ($I_{B1}$), as was the case in the $p^+$-$n$ diode. Hence, $I_E$ is approximately equal to the hole current injected into the base; that is, $I_{B1} \ll I_E$. Because $W_B$ is made $\ll L_p$, very few holes will be lost as they traverse the base and $I_{B2}$ will also be much less than $I_E$. Finally, because $I_{B3}$ corresponds to a reverse bias leakage current, it too will be very small and $I_C$ will be about equal to the hole current passing through the base. Adding these facts together, one concludes that for practical transistors under normal active region biasing conditions

1. The injected hole current from E–B is $\cong I_E$ with $I_E$ being slightly larger than $I_C$.

2. $I_B \ll I_C$ or $I_E$.

The large current gain capability of the bipolar transistor is depicted in Fig. 1.7 for a common emitter. A small base current provides electrons for recombination with holes coming from the emitter and for back injection into the emitter. The current gain $I_C/I_B$ is large because a $p^+$-$n$ junction (E–B) needs only a small electron current to provide a large hole current. Stated somewhat differently, a small base current forces the E–B to become forward biased and inject large numbers of holes.

## 1.3  FABRICATION

The fabrication of *pnp* and *npn* bipolar transistors is similar to that of *p-n* junction diodes as described in Volume II, Chapter 1. The major difference is that two *p-n* junctions must be formed for a *pnp* device—a $p^+$-$n$ junction for the E–B and a *n-p* junction for the C–B, with the base *n*-region being only a few microns in width.

Figures 1.8(a) and (b) illustrate a typical discrete, double-diffused *pnp* and an integrated circuit *npn* transistor, respectively. The discrete *pnp* is formed for starting with a $p^+$ (heavily doped) type substrate and growing a high resistivity, *p*-type epitaxial layer of silicon on the surface. The *p*-epitaxial layer is typically about 10 microns thick. The *n*-base region is thermally diffused through an oxide window. A $p^+$ (heavily doped) emitter is then diffused to form a $p^+np$ structure on the $p^+$-substrate. The collector contact on the bottom is made via the $p^+$-substrate. The geometrical arrangements of Fig. 1.8 are not to scale. To get a better idea of the size, stretch the figure 50 times in the horizontal direction, leaving the vertical dimensions fixed.

The integrated circuit *npn* transistor of Fig. 1.8(b) is fabricated starting with a high resistivity, *p*-type substrate. A small area of $n^+$ is diffused into the *p*-substrate and is called a buried layer. The primary function of the buried layer is to provide a low resistance path

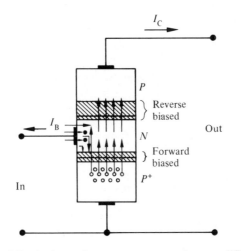

**Fig. 1.7**  Active region *pnp* common emitter amplifier.

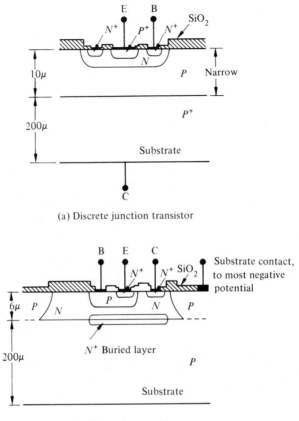

(a) Discrete junction transistor

(b) IC junction transistor

**Fig. 1.8**    (a) $p^+np$ discrete transistor; (b) $n^+pn$ integrated circuit transistor.

for the collector current. After the $n^+$ diffusion, a lightly doped $n$-type epitaxial layer is grown over the $n^+$ and $p$-substrate. The epitaxial $n$-region becomes the collector.

Isolation between components in an integrated circuit is achieved with the use of a reverse biased diode, called *diode isolation*. The $p^+$ regions are diffused through the $n$-epitaxial layer to make contact with the $p$-substrate. By placing the $p$-substrate at the most negative potential in the circuit, the $n$-epitaxial region is at some higher potential and therefore the $n$-epitaxial region is surrounded by a reverse biased $p$-$n$ junction.

The $p$-type base is diffused into the $n$-epitaxial region followed by the $n^+$ emitter and collector contact region. The collector contact at the surface is necessary in order to interconnect components on the surface, while the $n^+n$ collector contact is required because aluminum on $n^+$ forms an ohmic contact. Aluminum on lightly doped $n$-silicon can form a rectifying contact.

The remaining chapters assume a one-dimensional, bipolar analysis; that is, the major current flow is in only one direction. Because of the very narrow vertical dimension and little lateral current flow (except to the base contact), the assumption yields good first-order models. In particular, it gives equations that can be solved without reverting to numerical integrations.

## 1.4   CIRCUIT DEFINITIONS

The circuit variables necessary for the description of the bipolar transistor in the active region are presented in this section. Figure 1.9 illustrates the current components of the emitter and collector currents. The holes injected from E – B are $I_{Ep}$ and the electrons back injected from the base are $I_{En}$. Those holes injected from the emitter that reach the collector junction are $I_{Cp}$. The current $I_{Cn}$ is a result of the thermally generated electrons near the C – B junction that drift into the base. Equations (1.1) through (1.3) are the terminal currents written in terms of the active region current components.

$$I_E = I_{Ep} + I_{En} \tag{1.1}$$

$$I_C = I_{Cp} + I_{Cn} \tag{1.2}$$

$$I_B = I_E - I_C = I_{B1} + I_{B2} - I_{B3} \tag{1.3}$$

The *base transport factor* is defined as the ratio of the hole current diffusing into the collector to the hole current injected at the E – B junction, Eq. (1.4),

$$\alpha_T = I_{Cp}/I_{Ep} \tag{1.4}$$

Ideally $\alpha_T$ would be unity; however, with recombination of holes in the base $\alpha_T$ is slightly less than unity. In well fabricated devices $\alpha_T \to 1$.

Another performance parameter is the *emitter injection efficiency* ($\gamma$), which measures the injected hole current compared to total emitter current. Equation (1.5) is the definition,

$$\gamma = \frac{I_{Ep}}{I_E} = \frac{I_{Ep}}{I_{En} + I_{Ep}} \tag{1.5}$$

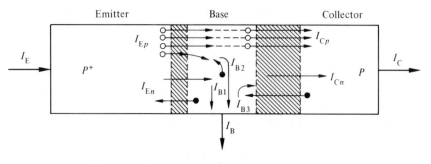

**Fig. 1.9**   Active region $p^+np$ current components.

Note that $\gamma \to 1$ if $I_{En} \to 0$; that is, as the emitter is more heavily doped, $I_{En}$ becomes a smaller percentage of $I_E$ (similar to the $p^+$-$n$ diode current components). For larger gain transistors, $\gamma$ is made as close to unity as possible.

The ratio of $I_C/I_E$ in the active region is defined as the *dc alpha* ($\alpha_{dc}$), and from Eqs. (1.1) and (1.2),

$$\alpha_{dc} = \frac{I_C}{I_E} = \frac{I_{Cp} + I_{Cn}}{I_{Ep} + I_{En}} \tag{1.6}$$

For almost any degree of E–B forward bias $I_{Cp} \gg I_{Cn}$ and Eq. (1.6) is approximated by Eq. (1.7),

$$\alpha_{dc} = \frac{I_{Cp}}{I_{Ep} + I_{En}} \tag{1.7}$$

If $I_{Cp}$ and $I_{Ep}$ are factored out of Eq. (1.7) and the definition of $\alpha_T$ is applied,

$$\alpha_{dc} = \frac{I_{Cp}}{I_{Ep}} \left[ \frac{1}{1 + I_{En}/I_{Ep}} \right] = \alpha_T \left[ \frac{I_{Ep}}{I_{Ep} + I_{En}} \right] \tag{1.8}$$

where, from the definition of $\gamma$,

$$\alpha_{dc} = \gamma \alpha_T \tag{1.9}$$

Ideally, as $\gamma$ and $\alpha_T$ approach unity, so will $\alpha_{dc}$. What is good for $\gamma$ and/or $\alpha_T$ is good for $\alpha_{dc}$.

Beta ($\beta_{dc}$) is another performance parameter for the active region and is defined by Eq. (1.10).

$$\beta_{dc} = \frac{I_C}{I_B} = \frac{I_C}{I_E - I_C} \tag{1.10}$$

By factoring $I_E$ from the denominator of Eq. (1.10),

$$\beta_{dc} = \frac{I_C}{I_E(1 - I_C/I_E)} = \frac{I_C/I_E}{1 - I_C/I_E} = \frac{\alpha_{dc}}{1 - \alpha_{dc}} \tag{1.11}$$

from Eq. 1.6, the definition of $\alpha_{dc}$. Note that, as $\alpha_{dc} \to 1$, $\beta_{dc} \to \infty$. Since $I_C/I_B$ is a current gain, we see the desirability for $\gamma$ and $\alpha_T$ (therefore $\alpha_{dc}$) to be as close to unity as possible.

The common base active region is most easily described by expanding Eq. (1.2) by using Eqs. (1.4) and (1.5).

$$I_C = I_{Cp} + I_{Cn} = \alpha_T I_{Ep} + I_{Cn} = \gamma \alpha_T \frac{I_{Ep}}{\gamma} + I_{Cn} = \alpha_{dc} I_E + I_{Cn} \tag{1.12}$$

The collector leakage current $I_{Cn}$ is defined by Eq. (1.13), as

$$I_{Cn} = I_{CB0} \tag{1.13}$$

the C–B current flowing with the emitter open circuited, similar to the reverse biased $n^+$-$p$

diode. The collector current is then obtained from Eq. (1.12) as

$$I_C = \alpha_{dc}I_E + I_{CB0} \tag{1.14}$$

The common emitter, active region, collector current can be obtained by substituting Eq. (1.3) into Eq. (1.14).

$$I_C = \alpha_{dc}(I_B + I_C) + I_{CB0} \tag{1.15}$$

Solving for $I_C$,

$$I_C(1 - \alpha_{dc}) = \alpha_{dc}I_B + I_{CB0} \tag{1.16}$$

$$I_C = \frac{\alpha_{dc}}{1 - \alpha_{dc}} I_B + \frac{I_{CB0}}{(1 - \alpha_{dc})} = \beta_{dc}I_B + I_{CB0}(\beta_{dc} + 1) \tag{1.17}$$

If the C–E leakage current ($I_{CE0}$) is defined as the case of $I_B = 0$, then from Eq. (1.17)

$$I_{CE0} = I_{CB0}(\beta_{dc} + 1) \tag{1.18}$$

and

$$I_C = \beta_{dc}I_B + I_{CE0} \tag{1.19}$$

We have now developed a set of equations for the active region operation for a *pnp* bipolar transistor. The task of Chapter 2 will be to derive the quantitative equations for $I_C$, $I_E$, and $I_B$ in terms of the material parameters such as $N_A$, $N_D$, $\tau_p$, etc., and in terms of the E–B and C–B voltages.

## PROBLEMS

**1.1** A $n^+pn$ bipolar transistor is at thermal equilibrium.

(a) Sketch the energy band diagram.

(b) Sketch the charge density and electric field.

(c) Sketch the potential using the $n^+$-region as $V = 0$.

(d) If in the active region, draw the particle flux and current diagram similar to Fig. 1.6.

**1.2** For the $n^+pn$ indicate the polarity of $V_{BE}$ and $V_{BC}$ for

(a) active region

(b) saturation region

(c) cutoff

(d) inverted active

(e) inverted saturation

**1.3** For a $p^+np$ device, indicate the polarity of $V_{EB}$ and $V_{CB}$ for

(a) saturation

(b) inverted active

(c) inverted saturation

(d) inverted cutoff

**1.4** For a *pnp* device $I_{Ep} = 1$ mA, $I_{En} = 0.01$ mA, $I_{Cp} = 0.98$ mA, and $I_{C_n} = 0.0001$ mA, calculate

(a) base transport factor

(b) emitter injection efficiency

(c) $\alpha_{dc}$ and $\beta_{dc}$; the value of $I_B$

(d) $I_{CBO}$ and $I_{CEO}$

(e) if $I_{Cp}$ were 0.99 mA, calculate $\beta_{dc}$ and $I_B$;

(f) if $I_{Cp} = 0.99$ mA and $I_{En} = .005$ mA, calculate $\beta_{dc}$ and $I_B$.

(g) how will $\beta_{dc}$ change if $I_{En}$ is increased?

# 2 / The Ideal Junction Transistor

The modern transistor has very narrow base region width and therefore little recombination in the base quasi-neutral bulk region. We define the *ideal bipolar junction transistor* as a device with no recombination–generation in the base bulk region, the E–B depletion region, and C–B depletion region. Also assumed is low-level injection, no electric fields in the bulk regions, one-dimensional current flow, and no external sources of generation such as light. Equations for the case of recombination in the base bulk region are presented without derivation.

The purpose of Chapter 2 is to use the ideal bipolar transistor model to develop quantitative expressions relating the terminal currents to the applied voltages between the E–B and B–C terminals. The derivation is preceded by a coherent "game plan" so that the reader has less chance of straying from the fold. Our game plan develops the necessary equations and discussions without going through the detailed mathematical calculations. Once the game plan is well in hand the individual derivations are developed starting with the solution for the minority carrier concentration in the base region. After the minority carrier concentrations in the bulk regions are obtained, they are used to obtain the terminal currents. Throughout the development the reader should be aware of the similarity to the diode derivation.

## 2.1  QUANTITATIVE ANALYSIS: GAME PLAN

To obtain the theoretical current voltage expressions with a minimum amount of confusion, let us assume the transistor to be constructed as shown in Fig. 2.1. Other assumptions which we will use either explicity or implicity are: (1) the structure is one-dimensional (all internal variables depend only on $x$); (2) $W$ is less than a minority carrier diffusion length; (3) recombination–generation currents arising in the depletion regions are negligible (the same assumption we used in establishing the ideal diode equation); (4) uniform doping density in the emitter, collector, and base; (5) low-level injection in all bulk regions. Note from Fig. 2.2 that the zero axis of the coordinates, $x = 0$, has been chosen on the base side of the E–B depletion region. $W$, as previously defined, is the quasi-neutral part of the total base width. The $x'$ and $x''$ axes for the emitter and collector regions are also chosen to start at the depletion region edges.

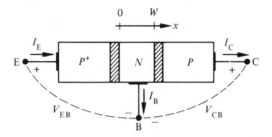

**Fig. 2.1** $p^+np$ current and voltage definitions.

### 2.1.1 Notation

The two $p$ regions of the $pnp$ transistor result in a notational problem. We are not able to use the same symbol for the electron diffusion length, $L_N$, in the emitter and the collector because with different doping $L_N$ has two different values. A generally accepted set of symbols is illustrated in Fig. 2.2, *The symbol refers to the minority carrier in the region*. For example, $L_N$ in the emitter becomes $L_E$, meaning the diffusion length for the minority carrier in the emitter, in this case for the $pnp$, the electron diffusion length. For the collector the symbol is $L_C$. For the base, $L_B$ is the minority carrier diffusion length in the base, for the $pnp$ this means $L_P = L_B$. Similar problems occur for the diffusion constants $D_N$ and $D_P$ as well as for the thermal equilibrium carrier concentrations $n_{p0}$ and $p_{n0}$. Figure 2.2 introduces the symbol $n_{E0}$ for the thermal equilibrium, minority carrier concentration in the emitter and $n_{C0}$ for the collector, while $D_E$ and $D_C$ are used for the electron diffusion constants. The base has $L_B$, $D_B$, and $p_{B0}$ for its minority carrier symbols. Note that the same set of symbols can be defined for the $npn$.

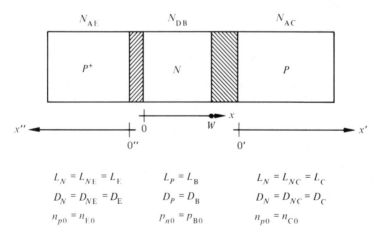

**Fig. 2.2** Definitions of coordinate axis and constants for the $pnp$.

### 2.1.2  Current Definitions

The ideal diode equation derivation used the minority carrier currents at the edges of the depletion region to determine the total current through the junction. Paralleling the ideal diode equation derivation, with no generation–recombination in the E–B or C–B depletion regions, the electron and hole currents entering a depletion region are constant throughout the region. Figure 2.3 illustrates the current components for the E–B and C–B depletion regions. The emitter current crossing the E–B depletion region is evaluated as two minority carrier diffusion currents, Eqs. (2.1) and (2.2), where the holes injected from the emitter into the base is the first term and that of the back-injected electrons from the base is the second.

$$I_E = I_{Ep}(0) + I_{En}(0'') \tag{2.1}$$

$$I_E = -qAD_B\frac{d\Delta p_B}{dx}\bigg|_{x=0} - qAD_E\frac{d\Delta n_E}{dx''}\bigg|_{x''=0} \tag{2.2}$$

Note that each term is evaluated at the edge of the depletion region. The minus sign of the second term of Eq. (2.2) is a result of the change of axis for electrons in the emitter. $I_E$ is positive when flowing in the negative $x''$ axis.

The collector current components at the C–B junction are determined by the holes entering the depletion region from the base side and by the electrons on the $p$-side. Again, each minority carrier diffusion current is evaluated at the depletion region edges; that is,

$$I_C = I_{Cp}(W) + I_{Cn}(0') \tag{2.3}$$

$$I_C = -qAD_B\frac{d\Delta p_B}{dx}\bigg|_{x=W} + qAD_C\frac{d\Delta n_C}{dx'}\bigg|_{x'=0} \tag{2.4}$$

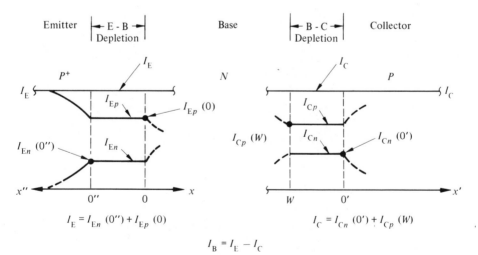

**Fig. 2.3**  Current components of $I_E$ and $I_C$ for a *pnp*.

The similarity of these definitions for junction transistor currents and the diode current derivation should be evident. If it is not, Chapter 3 of Volume II should be reviewed.

### 2.1.3  Bulk Region Solution

Equations (2.1) through (2.4) emphasize the need to know the minority carrier concentrations in each bulk region. The minority carrier concentrations $n_E(x'')$, $p_B(x)$, and $n_C(x')$ need to be derived. The solution to the minority carrier continuity equation assuming

1. there is low-level injection,
2. the electric field for the minority carrier is zero,
3. there is no generation due to light, etc., and
4. a steady-state, dc solution,

yields the minority carrier diffusion equations. Equation (2.5) is the equation to be solved for holes in the base region of a *pnp* device where $\tau_p = \tau_B$ by our notation.

$$D_B \frac{d^2 \Delta p_B(x)}{dx^2} = \frac{\Delta p_B(x)}{\tau_B} \tag{2.5}$$

Since the doping of the base is uniform, $p_{B0}$ is constant and any $x$ variation must be in $\Delta p_B$. Equation (2.5) is written as Eq. (2.6) by dividing both sides by $D_B$ and defining the diffusion length as $L_B = \sqrt{D_B \tau_B}$.

$$\frac{d^2 \Delta p_B(x)}{dx^2} = \frac{\Delta p_B(x)}{L_B^2} \tag{2.6}$$

$$\Delta p_B(x) = C_1 e^{x/L_B} + C_2 e^{-x/L_B} \tag{2.7}$$

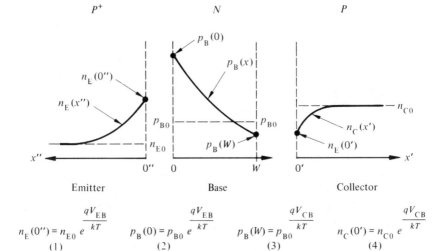

**Fig. 2.4**  Minority carrier concentrations and boundary conditions in the bulk regions.

Equation (2.7) is a solution to Eq. (2.6). Similar arguments result in diffusion equations for $\Delta n_E(x'')$ and $\Delta n_C(x')$, with $L_E = \sqrt{D_E \tau_E}$ and $L_C = \sqrt{D_C \tau_C}$. Each diffusion equation is a second-order differential equation and requires two boundary conditions for a complete solution.

## 2.1.4 Boundary Conditions

The junction voltages $V_{EB}$ and $V_{CB}$ establish the minority carrier concentration at the edges of the depletion regions. The derivation is identical to that for the $p$-$n$ junction diode and we present only the results in Eqs. (2.8) through (2.11).

$$\Delta n_E(0'') = n_{E0}(e^{qV_{EB}/kT} - 1) \tag{2.8}$$

$$\Delta p_B(0) = p_{B0}(e^{qV_{EB}/kT} - 1) \tag{2.9}$$

$$\Delta p_B(W) = p_{B0}(e^{qV_{CB}/kT} - 1) \tag{2.10}$$

$$\Delta n_C(0') = n_{C0}(e^{qV_{CB}/kT} - 1) \tag{2.11}$$

Figure 2.4 illustrates the carrier concentrations with the E–B forward biased and the C–B reverse biased, that is, active region operation.

## 2.1.5 Emitter Current

To solve for the emitter current $I_E$ we need to perform the following steps.

1. Solve the minority diffusion equation in the base, Eq. (2.6) for $\Delta p_B(x)$.
2. Apply the boundary conditions $\Delta p_B(0)$ and $\Delta p_B(W)$, Eqs. (2.9) and (2.10), to Eq. (2.7) and evaluate the two constants $C_1$ and $C_2$.
3. Solve the diffusion equation in the emitter bulk region, Eq. (2.12), for $n_E(x'')$

$$\frac{d^2 \Delta n_E(x'')}{dx''^2} = \frac{\Delta n_E(x'')}{L_E^2} \tag{2.12}$$

where $L_E = \sqrt{D_E \tau_E}$.

4. Apply the boundary conditions, namely $\Delta n_E(\infty) = 0$ and $n_E(0'')$, that is, Eq. (2.8).
5. Evaluate Eq. (2.2) for the diffusion current components at each edge of the depletion region as indicated in Fig. 2.3, which will yield $I_E$ as a function of $V_{EB}$ and $V_{CB}$.

## 2.1.6 Collector Current

To solve for the collector current $I_C$ the following steps are performed.

1. Solve a diffusion equation in the bulk collector region, Eq. (2.13), for $\Delta n_C(x')$.

$$\frac{d^2 \Delta n_C(x')}{dx'^2} = \frac{\Delta n_C(x')}{L_C^2} \tag{2.13}$$

where $L_C = \sqrt{D_C \tau_C}$.

2. Apply the boundary conditions, namely $\Delta n_C(\infty) = 0$ and $\Delta n_C(0')$, Eq. (2.11).
3. Evaluate Eq. (2.4) for the current components at each edge of the B–C depletion region, which will yield $I_C$ as a function of $V_{EB}$ and $V_{CB}$.

### 2.1.7  Base Current

To solve for $I_B$, apply Kirchhoff's Current Law, which states that

$$I_B = I_E - I_C \tag{2.14}$$

## 2.2  THE IDEAL BIPOLAR TRANSISTOR

As a special case of the bipolar transistor, we define the *ideal bipolar transistor* as having a base so narrow that no recombination occurs in the base. In modern devices the quasi-neutral base width $(W)$ is often less than 1 micron while the minority carrier diffusion length is about 35 microns. With such narrow base widths the minority carriers do not stay in the region long enough to have significant recombination. Therefore, we approximate the real device with the ideal case.

A second reason for starting with an ideal device is the greatly simplified equations that result for $I_E$ and $I_C$ where the current components can easily be identified. Several concepts can be easily demonstrated with the ideal model without getting lost in the mathematical detail.*

If no recombination of holes is to occur in the base region of a *pnp*, the lifetime $\tau_p = \tau_B$ must approach infinity; that is, $L_B$ must approach infinity, or alternatively, $W \ll L_B$. When applied to Eq. (2.6), the right side of the equation becomes zero.

$$\frac{d^2 \Delta p_B(x)}{dx^2} \cong 0 \tag{2.15}$$

Integrating twice and remembering the constants of integration yields Eq. (2.16),

$$\Delta p_B(x) = C_1 x + C_2 \tag{2.16}$$

Note that Eq. (2.16) is the equation of a straight line. Applying the two boundary conditions to Eq. (2.16) yields $C_2$ and then $C_1$.

$$\Delta p_B(0) = p_{B0}(e^{qV_{EB}/kT} - 1) = C_1[0] + C_2 \tag{2.17}$$

$$\Delta p_B(W) = p_{B0}(e^{qV_{CB}/kT} - 1) = C_1 W + C_2 \tag{2.18}$$

$$\Delta p_B(x) = \Delta p_B(0) - \left[\frac{\Delta p_B(0) - \Delta p_B(W)}{W}\right]x \tag{2.19}$$

Figure 2.5 illustrates the minority carrier concentrations for the ideal transistor in the

---

*It avoids writing $p_B(x)$ in terms of hyperbolic functions.

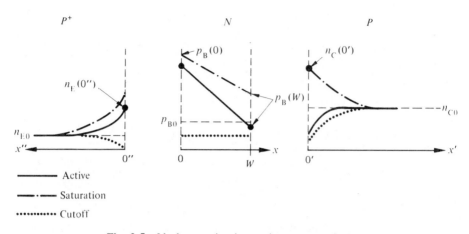

**Fig. 2.5**   Ideal *pnp* minority carrier concentrations.

active, saturation, and cutoff regions. Note that each case depends upon the polarities of $V_{EB}$ and $V_{CB}$.

A quick inspection of Fig. 2.5 and Eq. (2.19) indicates that the slope of $\Delta p_B(x)$ is

$$-\left[\frac{\Delta p_B(0) - \Delta p_B(W)}{W}\right] = \text{slope} \qquad (2.20)$$

which is used in calculating the hole current entering and leaving the base region.

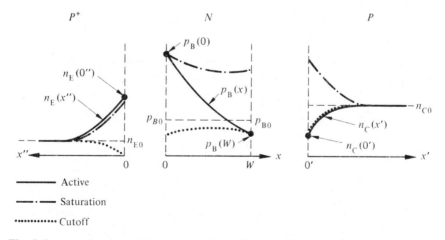

**Fig. 2.6**   *pnp* minority carrier concentrations with recombination in the base bulk region.

Since the slope is constant for the ideal device, the hole current is constant in the base region, which must be the case for no hole recombination. The hole particle flux into the base region equals the hole particle flux out of the base region. Comparing Fig. 2.5 with Fig. 2.6, the reader should note the difference in $\Delta p_B(x)$. For a device with significant recombination the plot is not a straight line; this indicates that holes are lost (recombine) in $W$.

## 2.3   THE IDEAL *pnp* TRANSISTOR CURRENTS

This section carries out the mathematical derivations of the ideal *pnp* bipolar transistor terminal currents as functions of the junction voltages. We are applying the "game plan" as outlined in Section 2.1 to the ideal device of Section 2.2.

To evaluate Eq. (2.2) for the emitter current, we need to first solve Eq. (2.12) for $\Delta n_E(x'')$ in the bulk emitter region. A similar equation was solved for the diode in Volume II, Chapter 3. The solution of the diffusion equation is Eq. (2.21).

$$\Delta n_E(x'') = C_1 e^{-x''/L_E} + C_2 e^{x''/L_E} \qquad (2.21)$$

Two boundary conditions are needed to solve for $C_1$ and $C_2$. Since an electron cannot survive forever in a $p$-region, $\Delta n_E(\infty) = 0$ and $C_2$ must be zero. At $x'' = 0''$,

$$\Delta n_E(0'') = C_1 e^{0''} = C_1 = n_{E0}(e^{qV_{EB}/kT} - 1) \qquad (2.22)$$

from Eq. (2.8) for the boundary condition at $x'' = 0''$. Therefore the solution is

$$\Delta n_E(x'') = n_{E0}(e^{qV_{EB}/kT} - 1)e^{-x''/L_E} \qquad (2.23)$$

Equation (2.23) is plotted in Figs. 2.4 and 2.5. The back injected electron current from the base to emitter, $I_{B1} = I_{En}(0'')$ (as sketched in Fig. 2.3), can now be evaluated from Eq. (2.2) as

$$I_{En}(0'') = -qAD_E\frac{d\Delta n_E(x'')}{dx}\bigg|_{x''=0''} = \frac{qAD_E}{L_E}n_{E0}(e^{qV_{EB}/kT} - 1) \qquad (2.24)$$

The hole current component $I_{Ep}(0)$ is easily evaluated for the ideal device because the slope has already been determined as Eq. (2.20).

$$I_{Ep}(0) = -qAD_B\frac{d\Delta p_B}{dx}\bigg|_{x=0} = \frac{qAD_B}{W}[\Delta p_B(0) - \Delta p_B(W)] \qquad (2.25)$$

Applying the boundary conditions, Eqs. (2.9) and (2.10), yields

$$I_{Ep}(0) = \frac{qAD_B}{W}p_{B0}[(e^{qV_{EB}/kT} - 1) - (e^{qV_{CB}/kT} - 1)] \qquad (2.26)$$

Equation (2.2) shows the emitter current to be the sum of Eqs. (2.24) and (2.26).

$$I_E = qA\left[\frac{D_E n_{E0}}{L_E} + \frac{D_B p_{B0}}{W}\right](e^{qV_{EB}/kT} - 1) - \frac{qAD_B}{W}p_{B0}(e^{qV_{CB}/kT} - 1) \qquad (2.27)$$

The collector current ($I_C$) can be derived by a similar set of calculations. However, we should be able to evaluate the first term of Eq. (2.4) directly. Since the hole current $I_{Ep}$ is a constant in the base (no hole recombination in the base), its value entering the base must be the same as that leaving the base; that is,

$$I_{Ep}(0) = I_{Cp}(W) \tag{2.28}$$

The current component $I_{Cn}(0')$ can be obtained directly from Eq. (2.24) by exchanging "E" for "C" and $V_{EB}$ for $V_{CB}$.* A change in sign is necessary because in this case $I_C$ is in the same direction as the $x'$ axis.

$$I_{Cn} = \frac{-qAD_C}{L_C} n_{C0}(e^{qV_{CB}/kT} - 1) \tag{2.29}$$

The collector current is the sum of Eq. (2.26) and (2.29).

$$I_C = \frac{qAD_B}{W} p_{B0}(e^{qV_{EB}/kT} - 1) - qA\left[\frac{D_C n_{C0}}{L_C} + \frac{D_B p_{B0}}{W}\right](e^{qV_{CB}/kT} - 1) \tag{2.30}$$

Equations (2.27) and (2.30) are applicable to all regions of operation, that is, active, saturation, cutoff, and inverted.

The base current for the ideal transistor is obtained by subtracting (2.30) from (2.27).

$$I_B = \frac{qAD_E}{L_E} n_{E0}(e^{qV_{EB}/kT} - 1) + \frac{qAD_C}{L_C} n_{C0}(e^{qV_{CB}/kT} - 1) \tag{2.31}$$

## 2.3.1   Active Region

Equations (2.27), (2.30), and (2.31) confirm the qualitative analysis of the active region *pnp* presented previously. For active-region operation $V_{CB}$ is negative and the exponential term is, for only a few tenths of a volt, much less than one, because $q/kT$ is about 40 at room temperature; that is,

$$\exp\left(\frac{qV_{CB}}{kT}\right) \ll 1, \quad \text{and} \quad \exp\left(\frac{qV_{EB}}{kT}\right) \gg 1 \tag{2.32}$$

because $V_{EB}$ is positive. The base current for the ideal device is primarily determined by the back-injection of electrons into the emitter, as discussed previously and by inspection of Eq. (2.31). The base current component $I_{B3}$, the second term of Eq. (2.31), represents the thermally generated electrons in the collector drifting from the edge of the collector to the base region, the leakage current of the C–B junction.

---

*Alternatively, Eq. (2.13) could be solved subject to the boundary conditions $\Delta n_C(\infty)$ and $\Delta n_C(0')$. Then apply the second term of Eq. (2.4) to get Eq. (2.29).

The collector current in the active region is dominated by the first term of Eq. (2.30), since for a $V_{EB}$ of only a few millivolts the exponential term becomes very large. The second term is small because $V_{CB}$ is negative and the "$-1$" dominates. The collector current is determined primarily by the holes injected from the emitter to base that are "collected" by the collector.

The emitter current Eq. (2.27) is also dominated by the $e^{qV_{EB}/kT}$ term. Because $p_{B0} \gg n_{E0}$, due to the heavily doped emitter compared to the base, and $L_E > W$, the hole current injected from the emitter to base is by far the largest emitter current component.

## 2.4  RECOMBINATION IN THE BASE

The minority carrier concentration in the base, $p_B(x)$, is altered from a straight line to a hyperbolic function when one permits recombination in the base region. A complete derivation is very similar to that of the previous sections with a different function for $p_B(x)$.

The derivation begins with the minority carrier diffusion equation in the base, Eq. (2.33a), which has Eq. (2.33b) as a solution.

$$\frac{d^2\Delta p_B(x)}{dx^2} = \frac{\Delta p_B(x)}{D_B \tau_B} = \frac{\Delta p_B(x)}{L_B^2} \qquad (2.33a)$$

$$\Delta p_B(x) = C_1 e^{-x/L_B} + C_2 e^{x/L_B} \qquad (2.33b)$$

The boundary conditions are determined by the junction voltages $V_{EB}$ and $V_{CB}$ (Eq. (2.9) and Eq. (2.10)). At $x = 0$,

$$\Delta p_B(0) = C_1 + C_2 = p_{B0}(e^{qV_{EB}/kT} - 1) \qquad (2.33c)$$

and at $x = W$,

$$\Delta p_B(W) = C_1 e^{-W/L_B} + C_2 e^{W/L_B} = p_{B0}(e^{qV_{CB}/kT} - 1) \qquad (2.33d)$$

Equations (2.33c) and (2.33d) must be solved simultaneously for the constants $C_1$ and $C_2$. The results are presented as Eq. (2.34).

$$\Delta p_B(x) = p_{B0}(e^{qV_{EB}/kT} - 1)\frac{\sinh\left(\dfrac{W-x}{L_B}\right)}{\sinh\left(\dfrac{W}{L_B}\right)} + p_{B0}(e^{qV_{CB}/kT} - 1)\frac{\sinh\left(\dfrac{x}{L_B}\right)}{\sinh\left(\dfrac{W}{L_B}\right)} \qquad (2.34)$$

Figure 2.6 illustrates the hole concentration for the active, saturation, and cutoff regions.

The emitter, collector, and base currents can now be derived for the case of recombination using the "game plan" of Section 2.1 and Eq. (2.34). The results are presented in Eq. (2.35a) through (2.35c).

$$I_E = \left[\frac{qAD_E}{L_E}n_{E0} + \frac{qAD_B}{L_B}p_{B0}\coth\left(\frac{W}{L_B}\right)\right](e^{qV_{EB}/kT} - 1) - \frac{qAD_B}{L_B}\frac{p_{B0}(e^{qV_{CB}/kT} - 1)}{\sinh\left(\dfrac{W}{L_B}\right)}$$

$$(2.35a)$$

$$I_C = \frac{qAD_B}{L_B}p_{B0}\frac{(e^{qV_{EB}/kT} - 1)}{\sinh\left(\dfrac{W}{L_B}\right)} - \left[\frac{qAD_Bp_{B0}}{L_B}\coth\left(\frac{W}{L_B}\right) + \frac{qAD_C}{L_C}n_{C0}\right](e^{qV_{CB}/kT} - 1)$$

$$\hspace{12cm}(2.35b)$$

$$I_B = I_E - I_C \hspace{5cm}(2.35c)$$

As one last parting comment, this set of equations can show that $I_{B2}$, the electrons furnished to the base for recombination with holes, is finite but small. However, such revelations from the present set of equations is not readily evident.

## 2.4.1  Quasi-ideal

A compromise between the ideal device (no recombination in $W$) and the general case, which results in rather abstract mathematical functions, is the quasi-ideal device. We define a *quasi-ideal device* as having a straight line minority carrier distribution in the base, as indicated by Eq. (2.19) and plotted in Fig. 2.5; yet it has recombination in the base ($I_{B2} \neq 0$). When $\Delta p_B(x)$ is plotted for a device with $W/L_B$ of about one-half (see Problem 2.2), the curve appears to approximate a straight line. Only when compared to a straight line is the difference readily apparent. Most modern day devices have $W/L_B$ of one-tenth or less, and therefore a straight line approximation in addition to recombination is reasonable.

The recombination current component of the base current, $I_{B2}$, is obtained by determining the total excess charge in the base which recombines on the average of $\tau_B$ seconds. With the aid of Eq. (2.19),

$$I_{B2} = \frac{\Delta Q_B}{\tau_B} = \frac{qA}{\tau_B}\int_0^W \Delta p_B(x)\,dx = \frac{qA}{\tau_B}\int_0^W \left[\Delta p_B(0) - \left(\frac{\Delta p_B(0) - \Delta p_B(W)}{W}\right)x\right]dx \quad (2.36)$$

$$I_{B2} = \frac{qA}{2\tau_B}[\Delta p_B(0) + \Delta p_B(W)]W = \frac{qAW}{2\tau_B}p_{B0}[(e^{qV_{EB}/kT} - 1) + (e^{qV_{CB}/kT} - 1)] \quad (2.37)$$

Note that the above equation is valid for all regions of operation and is simply added to Eq. (2.31), yielding an increased base current to

$$I_B = qA\left[\frac{D_E}{L_E}n_{E0} + \frac{W}{2\tau_B}p_{B0}\right](e^{qV_{EB}/kT} - 1) + qA\left[\frac{D_C}{L_C}n_{C0} + \frac{W}{2\tau_B}p_{B0}\right](e^{qV_{CB}/kT} - 1) \quad (2.38)$$

For typical devices, the first term of Eq. (2.38) may have $n_{E0} \sim 10^2$, $p_{B0} \sim 10^4$, $D_E \sim 10$, $L_E \sim 25~\mu$, $W \sim 1.5~\mu$ and $\tau_B \sim 1~\mu$, which yields

$$D_E n_{E0}/L_E \sim (10 \times 10^2)/25~\mu = 4 \times 10^5$$

while $Wp_{B0}/2\tau_B \sim 1.510^{-4} \times 10^4/2 \times 10^{-6} = 7.5 \times 10^5$ and indicates that $I_{B1}$ and $I_{B2}$ are of comparable size.

The emitter current is not significantly affected by recombination in the base for two reasons. First, $I_{B2}$, like $I_{B1}$, is much smaller than the injected hole current $I_{Ep}$ and $I_E \cong I_{Ep}$.

Secondly $I_{Ep}$ is determined by the slope of $p_B(x)$ at $x = 0$, which does not change from the ideal straight line case. We therefore do not change $I_E$ of Eq. (2.27) from the ideal case; that is, the emitter injection efficiency is unchanged.

The collector hole current $I_{Cp}$ is reduced because fewer holes arrive at the collector. Holes lost in recombination with electrons in the bulk base region create $I_{B2}$ and therefore reduce $I_C$ by that amount.

$$I_C = I_{C_{ideal}} - I_{B2} \tag{2.39}$$

From Eq. (2.30),

$$I_C = qA\left[\frac{D_B p_{B0}}{W} - \frac{W p_{B0}}{2\tau_B}\right](e^{qV_{EB}/kT} - 1)$$

$$- qA\left[\frac{D_C n_{C0}}{L_C} + \frac{D_B p_{B0}}{W} - \frac{p_{B0} W}{2\tau_B}\right](e^{qV_{CB}/kT} - 1) \tag{2.40}$$

Again note that

$$D_B/W \sim \frac{20}{10^{-4}} = 2 \times 10^5, \quad \text{and} \quad \frac{W}{2\tau_B} \sim \frac{10^{-4}}{2 \times 10^{-6}} = 50$$

and $I_{B2}$ will not change $I_C$ significantly in most cases. Therefore, the quasi-ideal device has the following.

$$I_E = I_{E_{ideal}} \quad \text{as Eq. (2.27)}$$
$$I_C = I_{C_{ideal}} \quad \text{as Eq. (2.30)}$$
$$I_B = I_{B_{ideal}} + I_{B2} \quad \text{as Eq. (2.38)}$$

The quasi-ideal bipolar transistor is a good compromise between simplicity of tracing the current components and the accuracy of the complete hyperbolic relationships. In any subsequent discussions, when we refer to $I_{B2} \neq 0$, we assume the relationship of Eq. (2.38).

## 2.5   IDEAL V–I CHARACTERISTICS

The current equations obtained in the previous sections can be used to plot the input and output characteristics for the common base and common emitter configurations. See Fig. 1.4 for the input and output variables.

### 2.5.1   Common Base

Figure 2.7(a) illustrates the input characteristics and Fig. 2.7(b) the output characteristics for the common base. Note that for $V_{CB} = 0$ the input characteristic is similar to that of a forward biased diode. Shorting the C–B and then using the E–B as a diode is commonplace in integrated circuits. For all $V_{CB}$ more negative than several $-kT/q$ volts, the input plot becomes one curve, indicating that the emitter current is

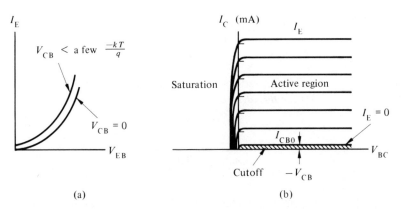

**Fig. 2.7**   Common base ideal *pnp* V–I characteristics: (a) input; (b) output.

primarily comprised of the forward injected holes (which are dependent on $V_{EB}$), independent of $V_{CB}$.

The common base output characteristics of Fig. 2.7(b) indicate that the active region
is to the right side of the $V_{BC}$ axis (or $V_{CB} < 0$). Note that $I_C$ is independent of $V_{BC}$ for
a fixed $I_E$. To the left of the axis, $V_{BC}$ is negative and the device is in saturation (for $I_E \geq 0$).
In saturation, holes are injected from the collector to the base; these holes are oppositely
directed to the holes being injected from the emitter. As a result, the collector hole current
components subtract from each other and the *net* collector current decreases. The more
forward biased the C–B junction becomes, the greater the collector injection of holes and
the smaller $I_C$. Note that for larger $I_E$, larger values of $V_{CB}$ are necessary for significant
hole current subtraction and therefore less $I_C$.

The cutoff region occurs when both the E–B and C–B junctions are reverse biased.
Figure 2.7(b) illustrates the case where $I_E = 0$ is the boundary between the active and
cutoff region.

The common base, active region, short circuit current gain $I_C/I_E$ is termed the *dc
alpha*, ($\alpha_{dc}$). The ratio of $I_C/I_E$, as defined from Eqs. (1.6) and (1.7), is obtained by
applying Eqs. (2.28), (2.26), and (2.24) *in the active region*.

$$\alpha_{dc} = \frac{\dfrac{D_B p_{B0}}{W}}{\dfrac{D_B p_{B0}}{W} + \dfrac{D_E n_{E0}}{L_E}} = \frac{1}{1 + \dfrac{D_E}{D_B}\dfrac{n_{E0}}{p_{B0}}\dfrac{W}{L_E}} = \frac{1}{1 + \dfrac{D_E}{D_B}\dfrac{N_{DB}}{N_{AE}}\dfrac{W}{L_E}} \qquad (2.41)$$

For the ideal device, alpha ($\alpha_{dc}$) is a measure of how few electrons are back injected
from the base to the emitter, compared with the holes injected from the emitter to the
base. As a performance criterion for the emitter current due to holes, as compared with
the total emitter current, the *emitter injection efficiency* ($\gamma$) was defined by Eq. (1.5) for

a *pnp* transistor.

$$\gamma = \frac{I_{Ep}}{I_E} = \frac{I_{Ep}}{I_{Ep} + I_{En}} \tag{2.42}$$

From Eqs. (2.26) and (2.27), the ratio of $I_{Ep}/I_E$, in the active region, is Eq. (2.43), which is identical to Eq. (2.41) *for the ideal device*.

$$\gamma = \frac{\dfrac{D_B p_{B0}}{W}}{\dfrac{D_B p_{B0}}{W} + \dfrac{D_E n_{E0}}{L_E}} = \alpha_{dc} \tag{2.43}$$

Note that in Eq. (2.43), as $n_{E0}$ is made much less than $p_{B0}$, by doping the emitter, $N_{AE} \gg N_{DB}$, that $\gamma \rightarrow 1$. If we could, we would make $\gamma = 1$, because this increases the gain of the transistor to phenomenal values ($\beta_{dc}$ would be infinite). The *base* transport factor ($\alpha_T$) is defined by Eq. (1.4); for the ideal device it is unity, because no holes are recombined (lost) as they travel through the base region.

The common base ideal *pnp* transistor output characteristics of Fig. 2.7(b) illustrate the definition of $I_{CB0}$ as the collector current, when the emitter is open-circuited ($I_E = 0$) and the C–B junction is reverse-biased. Physically, this means $I_{CB0}$ is the leakage current (reverse saturation current) of the C–B junction. For a $p^+np$ device, the leakage current is primarily determined by $n_{C0}$ of the collector because the base is doped more heavily than the collector; that is, $n_{C0} \gg p_{B0}$. The minority electrons thermally generated within one diffusion length of the C–B depletion region edge drift down the potential hill to the base and become $I_{B3}$ of the base current of Fig. 1.6. In cutoff $I_{B1}$ and $I_{B2}$ are zero since $I_E = 0$; that is, no holes are injected from the emitter and no electrons are back injected from the base.

## 2.5.2 Common Emitter

Figure 2.8(a) illustrates the input characteristics for the ideal common emitter *pnp* bipolar transistor. Note the similarity between the forward biased diode and the input characteristics, $I_B$ versus $V_{EB}$ with $V_{EC} = 0$. When $V_{EC}$ becomes greater than a few $kT/q$, the characteristic becomes independent of $V_{EC}$ because after the B–C junction becomes reverse biased by several $kT/q$, an additional voltage does not affect the injection of holes (and electrons) at the E–B junction. Figure 2.8(b) illustrates the output characteristic indicating the active region where, for a fixed $I_B$, $I_C$ becomes independent of $V_{EC}$ because further reverse bias on the B–C junction does not change the E–B voltage, which is fixed by $I_B$.

The active region is characterized by a large current gain. This current gain is called the dc beta ($\beta_{dc}$) and was defined as $I_C/I_B$ in Eq. (1.11). Because the ideal device in the active region has $V_{CB} < 0$ by several $kT/q$ volts, and because of the nonsymmetrical doping of the base and collector, $L_C \gg W$, and Eq. (2.30) can be approximated as Eq. (2.44) for $V_{EB} > 0.2$ volts.

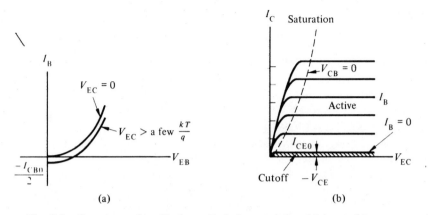

**Fig. 2.8**   Common emitter ideal *pnp* *V–I* characteristics: (a) input; (b) output.

$$I_C \cong qA\frac{D_B}{W}p_{B0}\,e^{qV_{EB}/kT} \tag{2.44}$$

For the base current a similar set of arguments yields Eq. (2.45).

$$I_B \cong qA\frac{D_E}{L_E}n_{E0}\,e^{qV_{EB}/kT} \tag{2.45}$$

The active-region beta ($\beta_{dc}$) is then the ratio of Eqs. (2.44) and (2.45).

$$\beta_{dc} = \frac{D_B}{D_E}\frac{p_{B0}}{n_{E0}}\frac{L_E}{W} = \frac{D_B}{D_E}\frac{N_{AE}}{N_{DB}}\frac{L_E}{W} \tag{2.46}$$

By doping the emitter heavily ($p^+$), $p_{B0} \gg n_{E0}$, and by making $W$ small, the current gain is made large. Again we see why the base region ($W$) is made very narrow.

If one lets $V_{CB} = 0$, the edge of the active region and saturation region, then the derivation above becomes exact. For $V_{CB} < 0$, the result is a good approximation because the C–B voltage does not significantly affect hole injection at the E–B junction.

The saturation region is defined as where both the E–B and C–B junctions are forward biased. As $V_{CB}$ becomes more forward biased, the number of *holes injected from the collector to the base increases*. The collector hole flux is opposite to the flux of holes arriving from the emitter; therefore the two components subtract. As illustrated in Fig. 2.8(b), the net collector current decreases as $V_{CB}$ increases (that is, $V_{EC}$ decreases) with a fixed value of $I_B$.

### 2.5.3   $I_{CE0}$

The leakage current from the emitter to the collector with the base open ($I_B = 0$) is defined as $I_{CE0}$. This is the collector current at the edge of cutoff, as illustrated in

Fig. 2.8(b). For $I_C < I_{CE0}$, the base current must be negative and the device is well into cutoff.

A physical explanation for why $I_{CE0} > I_{CB0}$ is based on the "transistor action" of the electrons in the base being back injected into the emitter. Figure 2.9 illustrates the currents and particle fluxes of the common emitter *pnp* with the base open circuited. The minority electrons within one diffusion length ($L_C$) of the C–B junction are swept through the depletion region of the reverse biased C–B junction. When they reach the base they are majority carriers and easily travel to the forward biased B–E junction. At the B–E junction the electrons become the source of electrons to be injected into the emitter. Because of the $p^+$ doping of the emitter, this relatively small number of electrons, originating in the collector, forces the injection of large numbers of holes into the base (as $\gamma \rightarrow 1$). For the ideal device with no recombination in the base, these holes reach the B–C junction and are collected by the collector. The total collector current is, as illustrated by Fig. 2.9, Eq. (2.47).

$$I_{CE0} = I_{Cn} + I_{Cp} \cong I_{CB0} + I_{Ep} \tag{2.47}$$

where $I_{Cn}$ is approximately the leakage current of the C–B junction with the emitter open, and $I_{Cp}$ is the injected holes from the emitter that reach the collector. Also note that, with no recombination in the base, $I_{CB0} = I_{En}$, the back injected electrons from base to emitter. From Eq. (1.18) repeated here for convenience,

$$I_{CE0} \cong I_{CB0}(\beta_{dc} + 1) \tag{2.48}$$

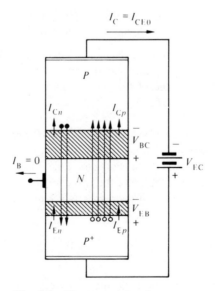

**Fig. 2.9**  The current $I_{CE0}$ for a *pnp*.

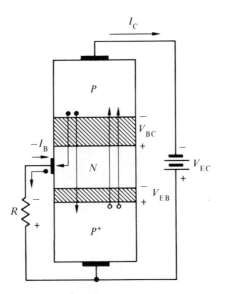

**Fig. 2.10**  $I_{CB0} \lesseqgtr I_C < I_{CB0}$ for a resistor between base and emitter.

To summarize, the collector current with the base open is much larger than just the leakage current of the reverse biased C–B junction because of the "transistor action" caused by the base current electrons forward biasing the E–B junction. Note the current gain of the device!

Figure 2.10 illustrates the case where $I_C$ is made less than $I_{CE0}$ by removing some of the electrons originating at the C–B junction, thereby preventing some of them from reaching the E–B junction. Note that the electrons extracted from the base result in a negative base current. In the case shown the forward biased E–B junction causes current to flow in the resistor $(R)$. The negative base current "steals" electrons from the base, thereby denying some back injected electrons to the emitter. With fewer back injected electrons the hole current from E–B is reduced, forcing $I_C$ to be less than $I_{CE0}$. In the limit as $R$ is reduced to zero (therefore $V_{EB} = 0$), $I_C$ becomes approximately $I_{CB0}$ and $I_B$ is approximately $-I_{CB0}$.

## 2.6  EBERS–MOLL EQUATIONS

Many of the computer-aided circuit analysis (CAD) programs written to solve transistor circuit problems use the nonlinear Eqs. (2.27), (2.30), and (2.31) to solve for the dc operating point variables. The computer must be used in their solution because of the nonlinear relationships between the currents and junction voltages. SPICE2 is one very often used CAD program for the analysis of bipolar as well as junction field effect and MOSFET transistor circuits.

The coefficients of Eq. (2.27) are given different names; for example, the first term is defined as

$$I_F = I_{F0}(e^{qV_{EB}/kT} - 1) = qA\left[\frac{D_E n_{E0}}{L_E} + \frac{D_B p_{B0}}{W}\right](e^{qV_{EB}/kT} - 1) \quad (2.49)$$

and the second term of Eq. (2.27) as

$$\alpha_R I_R = \alpha_R I_{R0}(e^{qV_{CB}/kT} - 1) = \frac{qA D_B}{W} p_{B0}(e^{qV_{CB}/kT} - 1) \quad (2.50)$$

Therefore the emitter current of Eq. (2.27) can be written as Eq. (2.51).

$$\boxed{I_E = I_F - \alpha_R I_R} \quad (2.51)$$

Similarly for the collector current of Eq. (2.30), the second term is defined as

$$I_R = I_{R0}(e^{qV_{CB}/kT} - 1) = qA\left[\frac{D_C n_{C0}}{L_C} + \frac{D_B p_{B0}}{W}\right](e^{qV_{CB}/kT} - 1) \quad (2.52)$$

and the first term as

$$\alpha_F I_F = \alpha_F I_{F0}(e^{qV_{EB}/kT} - 1) = \frac{qA D_B}{W} p_{B0}(e^{qV_{EB}/kT} - 1) \quad (2.53)$$

Therefore, the collector current of Eq. (2.30) is written as Eq. (2.54).

$$\boxed{I_C = \alpha_F I_F - I_R} \quad (2.54)$$

Since $I_B = I_E - I_C$, then Eq. (2.31) becomes

$$\boxed{I_B = (1 - \alpha_F)I_F + (1 - \alpha_R)I_R} \quad (2.55)$$

by subtracting Eq. (2.54) from Eq. (2.51). Equations (2.51), (2.54), and (2.55) are known as the *Ebers–Moll equations* for an ideal *pnp*. Figure 2.11 is the Ebers–Moll equivalent circuit. The circuit is just a restatement of the equations. Note the similarity of $I_F$ and $I_R$ to the E–B and C–B junction diode formulas. The current sources $\alpha_F I_F$ and $\alpha_R I_R$ illustrate the interaction terms of the two junctions due to the narrow base region.

The coefficients in front of the exponential terms on the right sides of Eqs. (2.50) and (2.53) are identical; therefore

$$\alpha_F I_{F0} = \alpha_R I_{R0} = I_S \quad (2.56)$$

where $I_S$ is defined by Eq. (2.56). The term $I_S$ is one of the required values for SPICE2. We can see that if

$$\beta_F = \frac{\alpha_F}{1 - \alpha_F} \quad (2.57)$$

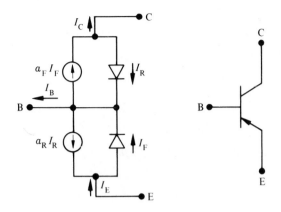

**Fig. 2.11**   Ebers–Moll equivalent circuit, *pnp*.

and

$$\beta_R = \frac{\alpha_R}{1 - \alpha_R} \tag{2.58}$$

then only three numbers are necessary for the Ebers–Moll equations to be completely specified; $\beta_F$, $\beta_R$, and $I_S$ (or $\alpha_F$, $\alpha_R$, and $I_S$). All the other parameters can be calculated.

The direct connection between the doping densities, base width, lifetimes, etc., and the Ebers–Moll equations makes them particularly attractive in integrated circuit analysis. Note that the complete set of input–output $V$–$I$ characteristics can be calculated from these equations. Also note that $\beta_F$ and $\beta_R$ are defined for all regions of operation and that in the active region $\beta_F \cong \beta_{dc}$.

By using the "complement" of a *pnp*, the *npn* has a set of Ebers–Moll equations similar to Eqs. (2.51), (2.54), and (2.55) and an equivalent circuit similar to Fig. 2.11 with the diodes and currents all reversed.

Several examples are presented to show the utility of the Ebers–Moll equations.

**Example:** $I_{CB0}$

For $I_E = 0$ and the B–C junction reverse biased, $I_C = I_{CB0}$. From Eq. (2.51) with $I_E = 0$

$$0 = I_F - \alpha_R I_R \cong I_F + \alpha_R I_{R0} \tag{2.59}$$

since $V_{CB}$ is large and negative, then $I_R \cong -I_{R0}$. Solving Eq. (2.59) for $I_F$ yields

$$I_F = -\alpha_R I_{R0} \tag{2.60}$$

Then for $I_E = 0$, substituting Eq. (2.60) into Eq. (2.54)

$$I_C = I_{CB0} = \alpha_F I_F - I_R = -\alpha_F \alpha_R I_{R0} + I_{R0} \tag{2.61}$$

and

$$I_{CB0} = (1 - \alpha_F \alpha_R) I_{R0} \tag{2.62}$$

**Example:** $I_{CE0}$

By definition $I_B = 0$ and the device is operated in the active region; that is

$$I_B = 0 = (1 - \alpha_F)I_F + (1 - \alpha_R)I_R \cong (1 - \alpha_F)I_F - (1 - \alpha_R)I_{R0} \qquad (2.63)$$

since $V_{CB}$ is large and negative, $I_R = -I_{R0}$, and

$$I_C = I_{CE0} = \alpha_F I_F + I_{R0} \qquad (2.64)$$

Solving Eq. (2.63) for $I_F$ and substituting it into Eq. (2.64) yields Eq. (2.65) for $I_{CE0}$

$$I_{CE0} = \frac{\alpha_F(1 - \alpha_R)I_{R0}}{(1 - \alpha_F)} + I_{R0} \qquad (2.65)$$

Further simplification of Eq. (2.65) results in Eq. (2.66)

$$I_{CE0} = I_{R0}\frac{(1 - \alpha_F\alpha_R)}{(1 - \alpha_F)} = I_{CB0}(\beta_F + 1) \qquad (2.66)$$

Equation (2.66) clearly illustrates the source of the carriers for $I_{CE0}$ and that $I_{CE0}$ is much larger than $I_{CB0}$.

The Ebers–Moll equations can also be obtained for the case of recombination in the base by defining the appropriate coefficients in front of the exponential terms of Eqs. (2.35a) through (2.35c).

## 2.7  SUMMARY

A "game plan" for the solution for a *pnp* transistor, with no generation or recombination in the E–B and C–B depletion regions, was developed to guide the reader through to a derivation of the terminal currents. The terminal currents were obtained by evaluating the minority carrier diffusion currents at the edge of each depletion region. Because no g–r occurs in the depletion region the currents must be constant throughout. Solving the minority carrier diffusion equations in the emitter, base, and collector with the appropriate boundary conditions resulted in the carrier distributions in each bulk region.

The ideal bipolar transistor was defined as a device having no recombination in the bulk base region; that is, the holes that enter must leave via the collector. The "game plan" as applied to the ideal *pnp* resulted in equations for $I_E$, $I_C$, and $I_B$, which were functions of $V_{CB}$ and $V_{EB}$. These three equations represent the dc nonlinear model for the device in all four regions of operation, that is, active, saturation, cutoff, and inverted operation.

A quasi-ideal device was defined as a compromise between the simple ideal case and the complex hyperbolic functions. With recombination in the base, an additional base current component is present and therefore models a real transistor.

The common base and common emitter $V$–$I$ characteristics were examined for the ideal device with alpha, beta, and the emitter-injection efficiency derived in terms of the material parameters. $I_{CB0}$ and $I_{CE0}$ were examined and their different magnitudes explained in terms of the transistor action. The chapter concludes with the Ebers–Moll equations.

## PROBLEMS

**2.1** Consider an ideal *pnp* device operating in the active region with the following parameters:

$n_{E0} = 2.56 \times 10^2/\text{cm}^3$    $L_B = 46.9 \times 10^{-4}$ cm    $L_C = 39.5 \times 10^{-4}$ cm

$L_E = 22.8 \times 10^{-4}$ cm    $D_B = 22$ cm$^2$/sec    $D_C = 15.6$ cm$^2$/sec

$D_E = 5.18$ cm$^2$/sec    $W = 4 \times 10^{-4}$ cm $= 4\ \mu$    $A = 1.265 \times 10^{-4}$ cm$^2$

$p_{B0} = 6.39 \times 10^3/\text{cm}^3$    $n_{C0} = 4.92 \times 10^5/\text{cm}^3$

$(N_{AE} \cong 3.9 \times 10^{17}/\text{cm}^3)$    $(N_{DB} \cong 1.57 \times 10^{16}/\text{cm}^3)$    $(N_{AC} = 2 \times 10^{14}/\text{cm}^3)$

Calculate at $V_{EB} = 0.67235$ and $V_{CB} = -1$

(a) The current component $I_{En}$ and $I_{Ep}$

(b) The current component $I_{Cp}$ and $I_{Cn}$

(c) $I_{B1}$ and $I_{B3}$

(d) How do the terms containing $V_{CB}$ compare with those of $V_{EB}$?

(e) What is the emitter injection efficiency, $\alpha_T$, and $\beta_{dc}$?

**2.2** Using the numerical data of $\Delta p_B(0) = 7.883 \times 10^{14}/\text{cm}^3$ and $\Delta p_B(W) = -6.39 \times 10^3/\text{cm}^3$, plot, on the same axis, $\Delta p_B(x)$ for the ideal device and the case of recombination in the base. For the sake of illustration let $W = 25.4\ \mu$ and $L_B = 46.9\ \mu$. What can be said about the slope of $\Delta p_B$ at $x = 0$ and $x = W$ as compared to the ideal? What about the area under the two plots?

**2.3** Using the data of Problem 2.1 as applied to the *quasi-ideal* device, calculate $I_{B2}$ and compare it to $I_{B1}$ and $I_{B3}$. Is base recombination significant?

**2.4** Calculate the Ebers–Moll coefficients for the ideal device of Problem 2.1, and examine the output *V–I* characteristic of the common emitter as the device is operated from active into deep saturation (with $I_B = 2\ \mu$A) by plotting $I_C$ versus $V_{EC}$ for three values of $V_{CB}$.

(a) $V_{CB} = -1$

(b) $V_{CB} = 0$ (edge of saturation-active)

(c) $V_{CB} = +.45$. What can be said about the amount of forward bias of $V_{CB}$ necessary to significantly reduce $I_C$ from its active region value?

**2.5** Explain how recombination in the base affects $I_{CE0}$ of the quasi-ideal device versus the ideal *pnp*.

**2.6** Using the concept of "complementary" devices, sketch the minority carrier concentrations for a $n^+pn$ device in the active and saturation region. Also make a sketch similar to Fig. 1.9 for each.

**2.7** Using data from Problems 2.1 and 2.3 calculate $\alpha_{dc}$ and $\beta_{dc}$ for the quasi-ideal device.

**2.8** If a $p^+np$ is operating in the inverted active mode, sketch all the minority carrier concentrations and a sketch similar to Fig. 1.9 for the current components and carrier flux, indicating the relative sizes.

**2.9** For a diffused base $p^+np$ the base region doping profile can be approximated by

$$N_{DB} = N_D(0)e^{-ax/W}, \qquad \text{where} \qquad a = \ln\left[\frac{N(0)}{N(W)}\right]$$

(a) Show that away from the edges of the base region the electric field is constant. Assume the device to be in thermal equilibrium. [*Hint*: $\mathscr{E} \propto a/W$.]

(b) Assume no recombination in the base and derive a formula for $p_B(x)$; let $p_B(W) = 0$ and $I_E \cong I_{Ep}$ the hole current in base.

(c) Calculate and plot $p_B(x)/(I_{Ep}W/qAD_{Ba})$ versus $x/W$ for $a = 5$ and $a = 9$.

(d) What can be surmised about the drift and diffusion components of hole current at $0.2\ W$, $0.5\ W$, and $0.9\ W$ for $a = 9$?

(e) Explain how this "graded base" will affect the chance of any recombination in the base.

# 3 / Deviations from the Ideal Transistor

The ideal bipolar transistor presented in Chapter 2 more than adequately describes most real devices in all their regions of normal operation. However, at very small currents and voltages or at large currents and voltages, the V–I characteristics of a real device, fabricated in silicon, deviate from the ideal device. The purpose of this chapter is to discuss the device physics of these deviations from the ideal and therefore more accurately describe a real bipolar transistor.

Comparisons between the ideal and observed V–I characteristics are shown in Figs. 3.1 and 3.2. Figure 3.1 displays common base characteristics while Fig. 3.2 displays common emitter characteristics. As can be seen from these figures, the basic shapes of the characteristics are about the same for the real and ideal devices. However, theoretically one predicts a single input characteristic for output voltages whose magnitudes are greater than a few $kT/q$ volts. In reality, the input characteristics are seen to change somewhat as the output voltage (either $V_{CE}$ or $V_{CB}$) is made more negative; that is, the C–B becomes more reverse biased. Likewise, the theoretical output characteristics are flat over most of the output voltage range. The observed output characteristics, on the other hand, slant upward with an increasing magnitude of the output voltage. All of the theoretical versus observed deviations noted above can be explained by taking into account *base width modulation*, also called the *Early effect*.

## 3.1 BASE WIDTH MODULATION

In drawing the theoretical ideal characteristics of Chapter 2 we implicitly assumed $\alpha_{dc}$, $\beta_{dc}$, $I_{C0}$, and $I_{CE0}$ to be constants independent of the applied biases. All of these quantities depend on or are functions of $W$. We have assumed that $W$, the quasi-neutral width of the base, was independent of bias. In truth, $W$ is not a constant independent of voltage. Changing $V_{EB}$ and/or $V_{CB}$ changes the depletion widths about the E–B and/or C–B junctions, thereby changing $W$ as illustrated in Fig. 3.3(a). The changes in the depleted portions of the base are especially significant because of the narrow physical extent of the base.

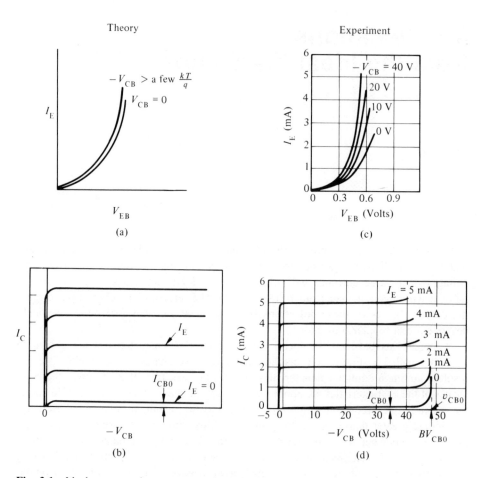

**Fig. 3.1**    Ideal versus real *pnp* common base devices: (a) ideal input; (b) ideal output; (c) real input (experimental); (d) real output (experimental).

The above effect, known as base width modulation or the Early effect (named for the engineer who first pointed out the phenomenon), can now be used to explain the observed variation in the input characteristics with output voltage, and the sloping portions of the output characteristics.

As seen from Eq. (2.27), repeated here as Eq. (3.1), which describes the common base input characteristics, $I_E$ is approximately proportional to $1/W$ and will therefore increase as $W$ is decreased due to a more negative value of $V_{CB}$.

$$I_E = qA\left[\frac{D_E n_{E0}}{L_E} + \frac{D_B p_{B0}}{W}\right](e^{qV_{EB}/kT} - 1) - \frac{qAD_B}{W}p_{B0}(e^{qV_{CB}/kT} - 1) \qquad (3.1)$$

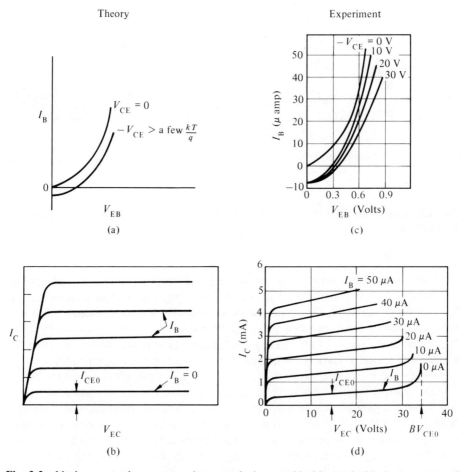

**Fig. 3.2** Ideal versus real common emitter *pnp* devices: (a) ideal input; (b) ideal output; (c) real input (experimental); (d) real output (experimental).

For $V_{CB} < -0.1$ volts and $V_{EB} > 0.1$ volts, Eq. (3.1) can be approximated by Eq. (3.2).

$$I_E \cong qA\left[\frac{D_E n_{E0}}{L_E} + \frac{D_B p_{B0}}{W}\right]e^{qV_{EB}/kT} \tag{3.2}$$

Also remember $p_{B0} \gg n_{E0}$ because the $p^+$ emitter is doped much greater than the base; hence

$$I_E \cong qA\frac{D_B p_{B0}}{W}e^{qV_{EB}/kT} \tag{3.3}$$

The base width $W$ clearly decreases as $V_{CB}$ becomes more negative, as shown in Fig. 3.3(b), and hence $I_E$ will be greater for a fixed $V_{EB}$ as illustrated in Fig. 3.1(c).

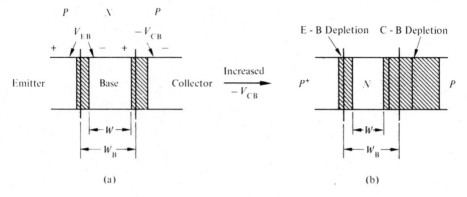

**Fig. 3.3**  Base width modulation effects as $-V_{CB}$ increases: (a) initial; (b) larger reverse bias of C–B junction.

The common base output characteristics of Fig. 3.1(d) show little base width modulation effects. Inspection of the active region shows only a very slight increase in $I_C$ as $V_{CB}$ changes for a fixed $I_E$. The phenomenon is somewhat self compensating because of the fixed emitter current condition. As $W$ becomes smaller $I_E$ wants to increase; therefore $V_{EB}$ decreases to keep $I_E$ fixed. The result is fewer holes injected into the base; they are the primary source of carriers for $I_C$; that is, $I_C$ will not increase significantly.

What slope exists in $I_C$ versus $V_{BC}$ is due to three possible phenomena. The first is a surface current around the reverse biased C–B junction, which in effect adds a small conductance in parallel with the junction. The second phenomenon occurs as $W$ shrinks. With a smaller $W$ there is even less of a chance that holes will recombine in the base, thereby increasing the number of injected holes that reach the collector; that is, an increase in $I_C$. This second phenomenon will occur only in devices where significant recombination is present in the base. The third is similar to the reverse biased junction diode where generation in the depletion region increased the reverse leakage current. As $V_{CB}$ becomes more negative, the C–B depletion region widens, allowing a larger generation current and thus an increase in $I_C$.

Figures 3.2(c) and (d) present the common emitter input and output characteristics that show larger effects of base width modulation. The Early effect on the actual device output characteristic shows a finite slope throughout the active region $I_C$ versus $V_{CE}$ plot (with a fixed $I_B$). Equation (2.31), for the ideal base current, shows that if $I_B$ is to be a constant in the active region, $V_{EB}$ must be constant, since by neglecting the last term (because $V_{CB}$ is negative),

$$I_B \cong \frac{qAD_E}{L_E}n_{E0}e^{qV_{EB}/kT} + I_{B2} \tag{3.4}$$

Similarly, the collector current of Eq. (2.30), in the active region where $V_{EC} > 2$ volts, can be approximated by Eq. (3.5),

$$I_C \cong \frac{qAD_B}{W}p_{B0}e^{qV_{EB}/kT} \tag{3.5}$$

For an increase in $V_{EC}$ ($V_{EB}$ = constant), $V_{CB}$ becomes more negative and the base width $W$ decreases causing an increase in $I_C$.

Another important phenomenon contributing to the slope of the common emitter output characteristics is the generation current produced in the C–B depletion region. Electrons and holes are generated, each contributing to an increase in $I_C$ as the C–B depletion region increases. However, the largest increase is a result of the generated electrons drifting into the base, where they become majority carriers. These electrons add directly to the back injected electrons at the E–B junction which forces a much larger increase in the injected holes from the emitter which in turn diffuse to the collector junction and increase $I_C$ significantly. Note that this device phenomenon is very similar to the $I_{CB0}$ and $I_{CE0}$ relationship. Figure 3.2(d), for $I_B$ = 0, shows the effect of the C–B depletion width increase (more carrier generation) as it increases $I_{CE0}$ with $V_{EC}$ increasing.

## 3.2  PUNCH-THROUGH AND AVALANCHING

### 3.2.1  Punch-Through

The term punch-through (or reach-through) refers to the physical situation where base width modulation has resulted in $W$ = 0; that is, the base is said to be punched-through when the E–B and C–B depletion regions touch each other inside the base as illustrated in Fig. 3.4. Once punch-through has occurred, the emitter-base and collector-base junctions become electrostatically coupled. Increases in $-V_{CB}$ beyond the punch-through point cause a lowering of the E–B junction potential hill, as illustrated in Fig. 3.5, allowing a large injection of holes from the emitter directly to the collector. Only a slight increase in $V_{BC}$ is necessary for a large increase in $I_C$. The effect of punch-through on the output characteristics may be seen in Figs. 3.1(d) and 3.2(d) as the rapidly increasing $I_C$ at large

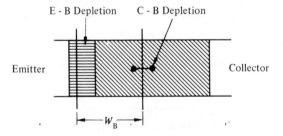

**Fig. 3.4**  Punch-through phenomenon.

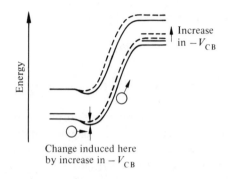

**Fig. 3.5**  Punch-through energy band diagram.

values of $-V_{CB}$ and $V_{EC}$, respectively. We say "may" because the $-V_{CB}$ required for punch-through must be greater than that required for another phenomenon giving rise to a similar high current effect, namely *avalanching*. The high-current regions are caused by either avalanching or punch-through, whichever occurs first.

### 3.2.2  Avalanche Breakdown

The collector to base junction, when reverse biased, is similar to the diode, and avalanching will cause a large reverse current to flow if the reverse voltage across a junction approaches the value at which the electrons and holes, as they drift in the C–B depletion region, gain sufficient energy to ionize silicon atoms and generate additional electron–hole pairs. Under active region operation and a progressively increasing $-V_{CB}$, a point is eventually reached where the collector-base junction begins avalanching (assuming, of course, that punch-through has not occurred). Based on this line of reasoning alone, one would expect output characteristics of the form shown in Fig. 3.6(a) for the common base. The *collector to base breakdown voltage* $BV_{CB0}$ is defined as the B–C voltage where $I_C$ increases very rapidly, $I_C \to \infty$ (typically a factor of 10 to 100 over the low-voltage value) with $I_E = 0$. Note that an increase in $I_C$ occurs long before $BV_{CB0}$. Instead of thinking of avalanching as occurring at one specific voltage, we must recognize that in reality, a degree of avalanching occurs long before what is normally designated as the breakdown voltage. In fact, some avalanching occurs before any increase in $I_C$ is perceived on the $I_C$ versus $-V_{CB}$ plot.

The common emitter avalanching characteristic is illustrated in Fig. 3.6(b) and characterized by the *collector to emitter breakdown voltage* $BV_{CE0}$. First note that $BV_{CE0}$ is much less than $BV_{CB0}$, and secondly note that the avalanching is more spread out. The breakdown voltage $BV_{CE0}$ is defined with the base open-circuited; that is, $I_B = 0$. A small amount of avalanching at the C–B junction produces additional electrons that enter the base region, drift through the base (as majority carriers), and supply the additional back injected electrons to the emitter. The additional back injected electrons force a large

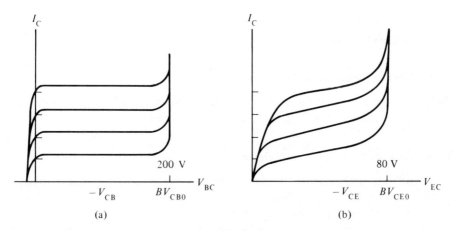

**Fig. 3.6** *pnp* avalanching: (a) common base; (b) common emitter.

increase in the number of holes injected from the emitter to the base. These injected holes diffuse to the collector and add a large component to the collector current. Effectively the current from the avalanching electrons is multiplied by the "transistor action" of the device. Therefore, even at low values of avalanche multiplication, a significant increase in $I_C$ is observed on the output characteristics of the common emitter. Because of the "transistor action" of the electrons reaching the emitter, $BV_{CE0}$ is much less than $BV_{CB0}$. The reader should note the similarity to the $I_{CB0}$ versus $I_{CE0}$ arguments presented in Chapter 2.

As a final observation, note that any resistor placed between the base and emitter will increase the breakdown voltage for the common emitter. The E–B resistor will "steal" some of the avalanching electrons, prohibiting them from reaching the emitter. With fewer electrons for back injection to the emitter, fewer holes are injected into the base and contribute to $I_C$. The value of $R$ controls the breakdown voltage as indicated by Eq. (3.6).

$$BV_{CB0} > BV_{CER} > BV_{CE0} \tag{3.6}$$
$$(R = 0) \qquad\qquad (R = \infty)$$

Figures 2.9 and 2.10 illustrate the effect of reducing the number of electrons reaching the emitter by draining them out through the base terminal. In our present case the extra electrons are created in the C–B depletion region by avalanching. As $R$ decreases, more electrons are removed from the base region.

## 3.3 GEOMETRY EFFECTS

In our ideal model for the bipolar transistor it was assumed that the device was one-dimensional. Practical transistors are always three-dimensional. Figure 1.8 clearly shows the two-dimensional cross sections of a discrete planar and an integrated circuit device.

The three- (or two)-dimensional nature of a practical transistor necessitates a revision of the ideal transistor to take into account several factors.

### 3.3.1 Emitter Area ≠ Collector Area

The emitter area is not equal to the collector area, as assumed in the simple model. This is illustrated in Fig. 3.7(a) using the discrete planar transistor.

### 3.3.2 Bulk and Contact Resistance

The base current entering the base terminal must pass through a resistive region before it reaches the "heart" of the transistor (dotted lines in Fig. 3.7(b)). Therefore, the voltage drop across the junction is actually somewhat less than the terminal voltage $V_{EB}$. (This can be quite serious since the E–B junction is forward biased under normal active mode operation.) Also note that the $n^+$ base contact has a metal-to-semiconductor junction. This junction is usually ohmic (contact resistance) and contributes to the overall base series resistance $r_B$. Figure 3.7(a) illustrates the bulk collector resistance, $r_C$, which can be significant as the collector is always lightly doped.

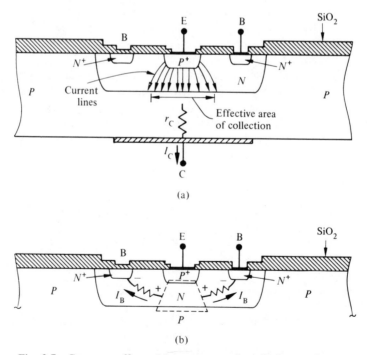

**Fig. 3.7** Geometry effects: (a) current crowding; (b) base resistance.

To reduce $r_C$ the planar device is made on a thin ($10~\mu$) epitaxial layer on a $p^+$ substrate as illustrated by Fig. 1.8(a). The integrated circuit device reduces $r_C$ by placing the "buried layer" of $n^+$ material below the collector region. The buried layer provides a low resistance path to the collector contact as shown by Fig. 1.8(b). Even though the emitters are formed on heavily doped material, the metal-semiconductor contact forms a series resistance $r_E$.

### 3.3.3  Current Crowding

In addition to a voltage drop in getting to the heart of the transistor there will also be a voltage drop across the face of the emitter due to the distributed base resistance. The center of the emitter will be less forward biased than the outer edges. This leads to a larger current around the edges than in the middle of the emitter, a situation known as *current crowding*. Figure 3.7(a) illustrates this case. It is for these reasons that power transistors are designed to have a large emitter perimeter-to-area ratio and several base contacts. Often interdigitated emitters and base contacts form a comb structure by connecting 10 to 20 emitters in parallel.

## 3.4  GENERATION–RECOMBINATION (g–r) IN THE DEPLETION REGION

The ideal transistor was defined as having no generation or recombination in the E–B and C–B depletion regions. As in the diode, a reverse biased junction with generation in its depletion region will add a generation current component to the leakage current. For example, $I_{CB0}$ will be increased as electrons and holes are generated in the C–B depletion region and fall down the potential hill. Also note that the generation component of the leakage current will increase as $|V_{CB}|^{1/2}$ (for an abrupt C–B junction). The C–B generation current will also increase $I_{CE0}$ due to the additional electrons available for back injection at the E–B junction.

A forward biased junction will have recombination in its depletion region and therefore an additional current component to the emitter and base current. For a device in the active region at low values of $I_B$, Fig. 3.8 illustrates the effect on $I_B$. Because $I_{En} \cong I_B$ and $I_{En} \ll I_E$, the generation current affects $I_B$ much more than $I_E$. The arguments for explaining the recombination current in the E–B depletion region are identical to those for the diode. Also plotted in Fig. 3.8 is the collector current, which is not affected by the E–B generation–recombination currents because $I_C$ is primarily due to those holes injected into the base that diffuse to the collector.

The $\beta_{dc}$ of the transistor is the ratio of $I_C$ to $I_B$ in the active region. By plotting $\ln I_C$ versus $V_{EB}$, Eq. (3.5), on the same scale as $\ln I_B$ in Fig. 3.8, we find the difference between the plots to be $\ln \beta_{dc}$. Figure 3.9 is $\beta_{dc}$ plotted over many decades of collector current as transcribed from Fig. 3.8. Note that at low current levels the recombination current in the E–B depletion region causes a reduction in $\beta_{dc}$. As $I_C$ is increased, the recombination current becomes a smaller part of the total current injected and $\beta_{dc}$ increases.

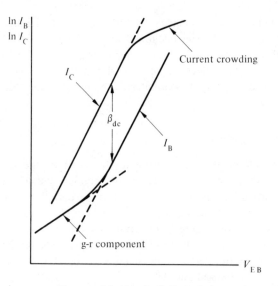

**Fig. 3.8** Active-region *pnp* with recombination in E–B depletion region and current crowding (or high-level injection).

### 3.4.1 Current Crowding Revisited

At the larger $I_C$ levels of Fig. 3.8, the effects of current crowding and/or high-level injection are illustrated. Because the collector current is comprised primarily of the holes injected at the E–B junction, and the E–B is much like the forward biased diode, $I_C$ will not follow the ideal equation forever. At a certain current density level we violate the low-level injection assumption and, as in the diode, the collector current in the active

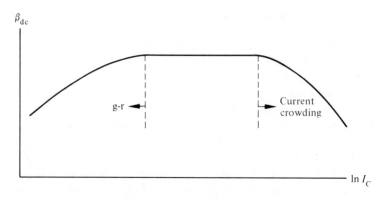

**Fig. 3.9** Nonideal effects on $\beta_{dc}$.

region becomes

$$I_C \cong \frac{qAD_B}{W} e^{qV_{EB}/nkT} \tag{3.7}$$

where $n \cong 2$. The condition for high-level injection is aggravated by the "current crowding" at the edges of the emitter; that is, because of the lateral base resistance the edges become more forward biased, leading to the high-level injection condition at lower current levels. The $\beta_{dc}$ "falloff" with large collector current is illustrated in Fig. 3.9.

## 3.5  SUMMARY

This chapter has presented several phenomena that explain the deviations from ideal as observed in real bipolar transistors. The Early effect, or base width modulation, explained why the output characteristics have a finite slope as $|V_{CB}|$ is increased. Punch-through and/or avalanching affect the large voltage output characteristics where $I_C$ increases rapidly. The differences between $BV_{CB0}$ and $BV_{CE0}$ were explained in terms of the transistor action of the avalanched C–B carriers. Geometry effects, contact, and bulk resistances and how they affect current flow were discussed. Generation and recombination in the depletion regions were used to explain the $\beta_{dc}$ versus $I_C$ changes at low levels of current. High-level injection at large values of $I_C$ results in the reduction of $\beta_{dc}$.

## PROBLEMS

**3.1** An ideal $p^+np$ device in the active region has $I_E = 961.3 \ \mu A$ and $W = 3 \ \mu$, when $V_{EB}$ is 0.65 volt.

(a) If base-width modulation effects are included, what is the rate of change of $I_E$ with respect to $W$?

(b) If $W$ is determined only by $W_B - x_n$ where $x_n \cong K(-V_{CB})^{1/2}$, what is the equation for the rate of change of $I_E$ with respect to $V_{CB}$?

(c) For fixed $V_{EB}$ determine an equation for the rate at which $I_B$ changes with respect to $V_{EC}$ if quasi-ideal.

**3.2** Use a sketch similar to Fig. 2.10 for an $n^+pn$ device to explain the difference between $BV_{CE0}$ and $BV_{CER}$.

**3.3** If recombination in the base bulk region were included in Figs. 3.8 and 3.9, sketch the new curve with respect to the old.

**3.4** Active region, base width modulation effects provide a slope to the active region $I_C - V_{EC}$ plot with $V_{EB}$ = constant. Derive a formula for $dI_C/dV_{EC}$, assuming an abrupt B–C junction and that $|V_{CB}| \gg V_{bi}$.

# 4 / Small Signal Models

The previous two chapters described the response of the bipolar junction transistor to dc voltages and currents. The present chapter will investigate the response of the transistor to a small signal voltage or current superimposed upon the dc values. The term "small signal" implies that the peak values of the signal current and voltage are much smaller than the dc values. Typically, this means a signal voltage of several millivolts or less.

Many small signal circuit models have been developed to represent the signal response. One particularly useful representation is called the *hybrid-pi model*. This model has several advantages that make it particularly attractive to circuit design engineers. For example, the model explicitly relates the signal model circuit element values to the dc operating point variables. Also, the temperature variations in the model parameters are easily obtained and for frequencies of typically less than 500 MHz, the circuit elements are frequency independent.

Small signal models are used for calculating signal gains, input impedances, and output impedances for amplifiers. Because the largest signal gain and least distortion are obtained from bipolar transistors operating in their active region, only active region models are particularly useful. We assume throughout this chapter that the transistor is operating in the active region.

## 4.1 LOW-FREQUENCY MODEL

The development of a low-frequency model for the bipolar transistor begins with the assumption of a quasi-static response of the carriers to small changes in the terminal voltages. By quasi-static we mean that the electrons and holes return to near steady state in a time much less than the period of the signal. Figure 4.1 illustrates the carrier response to a positive increment of emitter-to-base voltage, $\Delta V_{EB}$, with a period sufficiently long to allow the holes injected from the emitter to base and the electrons injected from base to emitter to reach steady state. The figure shows that in this case no signal voltage was applied to the C–B junction. Note that with a larger slope to $p_B(x)$, the emitter current

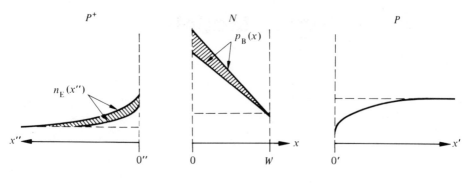

$$p_B(0) = p_{B0}\, e^{\,q(V_{EB} + \Delta V_{EB})/kT}$$

**Fig. 4.1**  Quasi-static response to $V_{eb}$.

has increased in response to $\Delta V_{EB}$; that is, the total instantaneous value $i_E$ has increased over the dc value $I_E$ by the signal component, where $\Delta V_{EB} \equiv v_{eb}$ by definition. Equation (4.1) is a formal statement of the emitter current components.

$$i_E = I_E + i_e \tag{4.1}$$

Similarly, the other voltages and currents are written as

$$v_{EB} = V_{EB} + v_{eb} \tag{4.2}$$

$$i_B = I_B + i_b \tag{4.3}$$

$$i_C = I_C + i_c \tag{4.4}$$

$$v_{CB} = V_{CB} + v_{cb} \tag{4.5}$$

Our goal is to develop relationships between $i_b$ and $i_c$ in terms of $v_{cb}$ and $v_{cc}$. This choice of variables is somewhat arbitrary but is most useful because it models the common emitter amplifier, which has $i_b$ as the input current and $i_c$ as the output current with $v_{cb}$ the input voltage and $v_{cc}$ as the output voltage.

The Ebers–Moll relationships are a convenient starting point, provided we assume quasi-static responses of the carriers in each bulk region. For low-frequency signal analysis we replace the dc variables in the Ebers–Moll equations (Eqs. (2.51), (2.54), and (2.55)) with the total instantaneous variables of Eqs. (4.1) through (4.5). Before making these substitutions, remember that we are looking for signal models useful for active region operation. Therefore, we simplify the Ebers–Moll equations to active region equations; that is, the E–B is forward biased and the C–B is reverse biased, and $e^{qV_{EB}/kT} \gg 1$ with $e^{qV_{CB}/kT} \ll 1$. Equations (2.51), (2.54), and (2.55) are approximated as

$$i_E = I_{F0}(e^{qv_{EB}/kT} - 1) - \alpha_R I_{R0}(e^{qv_{CB}/kT} - 1) \cong I_{F0}e^{qv_{EB}/kT} \tag{4.6}$$

$$i_C = \alpha_F I_{F0}(e^{qv_{EB}/kT} - 1) - I_{R0}(e^{qv_{CB}/kT} - 1) \cong \alpha_F I_{F0}e^{qv_{EB}/kT} \tag{4.7}$$

$$i_B = (1 - \alpha_F)I_{F0}(e^{qv_{EB}/kT} - 1) + (1 - \alpha_R)I_{R0}(e^{qv_{CB}/kT} - 1) \cong (1 - \alpha_F)I_{F0}e^{qv_{EB}/kT} \tag{4.8}$$

where

$$I_{FO} = qA \left[ \frac{D_E n_{EO}}{L_E} + \frac{D_E p_{BO}}{W} \right] \tag{4.9}$$

and

$$v_{CB} = v_{EB} - v_{EC} \tag{4.10}$$

The general forms of Eqs. (4.7) and (4.8), upon substituting Eqs. (4.1) through (4.5) and making use of Eq. (4.10) to replace $V_{CB}$ with $V_{EC}$, are of the form

$$i_C = f_1(V_{EB} + v_{eb}, V_{EC} + v_{ec}) = I_C + i_c \tag{4.11}$$

$$i_B = f_2(V_{EB} + v_{eb}, V_{EC} + v_{ec}) = I_B + i_b \tag{4.12}$$

If $v_{eb}$ and $v_{ec}$ are much smaller than $V_{EB}$ and $V_{EC}$, respectively, then the terms up to order $\Delta$ in the Taylor series expansion of Eqs. (4.11) and (4.12) are good approximations to the total instantaneous values. Mathematically the Taylor series is stated by Eq. (4.13).

$$f(x + \Delta x, y + \Delta y) = f(x, y) + \left.\frac{\partial f}{\partial x}\right|_y \Delta x + \left.\frac{\partial f}{\partial y}\right|_x \Delta y + \cdots \tag{4.13}$$

Let's expand Eq. (4.7) for the collector current, $i_C$, where $f_1$ is the functional form of Eq. (4.7) with $V_{EB}$ replaced by $V_{EB} + v_{eb}$ and $V_{EC}$ by $V_{EC} + v_{ec}$.

$$i_C = \underbrace{f_1(V_{EB}, V_{EC})}_{I_C} + \underbrace{\left.\frac{\partial f_1}{\partial v_{EB}}\right|_{V_{EC}} v_{eb} + \left.\frac{\partial f_1}{\partial v_{EC}}\right|_{V_{EB}} v_{ec}}_{i_c} \tag{4.14}$$

For low-frequency, small signal operation

$$\frac{\partial f_1}{\partial v_{EB}} \cong \frac{\partial f_1}{\partial V_{EB}} \quad \text{and} \quad \frac{\partial f_1}{\partial v_{EC}} \cong \frac{\partial f_1}{\partial V_{EC}}$$

After inspecting Eq. (4.14), we can write the signal collector current

$$i_c = \left.\frac{\partial f_1}{\partial V_{EB}}\right|_{V_{EC}} v_{eb} + \left.\frac{\partial f_1}{\partial V_{EC}}\right|_{V_{EB}} v_{ec} = g_m v_{eb} + g_0 v_{ec} \tag{4.15}$$

recalling that the function $f_1$ is that of a current. Therefore the incremental conductances are defined as $g_m$ the *transconductance*, $i_c/v_{eb}$ with $v_{ec} = 0$, and $g_0$ the *output conductance*, $g_0 = i_c/v_{ec}$, with $v_{eb} = 0$. The output resistance, with the input *signal* short-circuited, is $1/g_0 = r_0$.

Expanding Eq. (4.8) in a Taylor series expansion, with $f_2$ representing the functional form of the base current, yields

$$i_B = f_2(V_{EB}, V_{EC}) + \left.\frac{\partial f_2}{\partial v_{EB}}\right|_{V_{EC}} v_{eb} + \left.\frac{\partial f_2}{\partial v_{EC}}\right|_{V_{EB}} v_{ec} = I_B + i_b \tag{4.16}$$

where

$$i_b = \left.\frac{\partial f_2}{\partial v_{EB}}\right|_{V_{EC}} v_{eb} + \left.\frac{\partial f_2}{\partial v_{EC}}\right|_{V_{EB}} v_{ec} = (g_\pi + g_\mu)v_{eb} - g_\mu v_{ec} \tag{4.17}$$

The incremental signal conductances are defined as $g_\pi + g_\mu = i_b/v_{eb}$, the *input* conductance with the output signal $v_{ec} = 0$, and the *reverse feedback conductance* $g_\mu = -i_b/v_{ec}$ with $v_{eb} = 0$.

## 4.2  LOW-FREQUENCY HYBRID-PI MODEL

The most popular of many possible low-frequency signal models for the bipolar transistor, is the hybrid-pi. Figure 4.2(a) illustrates the simple, ideal, low-frequency device model. If we assume an ideal bipolar transistor with no base width modulation, then $W$ is a constant in Eq. (4.9) and $I_{F0}$ is a constant independent of $v_{EB}$ and $v_{EC}$. Applying Eq. (4.15) to Eq. (4.7) and using Eq. (4.6) for $I_E$ we find $g_m$ and $g_0$,

$$g_m = \left.\frac{\partial I_C}{\partial V_{EB}}\right|_{V_{EC}} = \alpha_F I_{F0}\frac{q}{kT}e^{qV_{EB}/kT} = \alpha_F\frac{q}{kT}I_E = \frac{q}{kT}I_C \tag{4.18}$$

$$g_0 = \left.\frac{\partial I_C}{\partial V_{EC}}\right|_{V_{EB}} = 0 \tag{4.19}$$

Note the absence of $g_0$ from Fig. 4.2(a). A similar application of Eq. (4.17) to Eq. (4.8) yields $g_\pi$ and $g_\mu$.

$$g_\pi = \frac{1}{r_\pi} = \left.\frac{\partial I_B}{\partial V_{EB}}\right|_{V_{EC}} = (1 - \alpha_F)I_{F0}\frac{q}{kT}e^{qV_{EB}/kT} = \frac{q(1 - \alpha_F)}{kT}I_E \tag{4.20}$$

$$r_\pi = \frac{kT}{q(1 - \alpha_F)I_E} = \frac{kT\alpha_F}{q(1 - \alpha_F)\alpha_F I_E} = \frac{kT\beta_F}{qI_C} = \frac{\beta_F}{g_m} \tag{4.21}$$

$$g_\mu = \left.\frac{\partial I_B}{\partial V_{EC}}\right|_{V_{EB}} = 0 \tag{4.22}$$

Note that this model has a perfect (ideal) signal transmission path from input to output with no reverse feedback, that is, a unilateral signal path. The parameter $\beta_F$ is approximately the same as $\beta_{ac}$ in the active region and is sometimes written as $\beta_0$, the *low-frequency signal beta* of the device.

### 4.2.1  Nonideal

A bipolar transistor that has base width modulation will have the small signal equivalent circuit of Fig. 4.2(b). In this case $W$ is a function of $v_{EB}$ and $v_{EC}$ (by way of $v_{CB}$), and $g_\mu$ and $g_0$ are not zero. In most practical cases $g_\mu$ is at least 100 times smaller than $g_0$ and is often ignored in circuit applications.

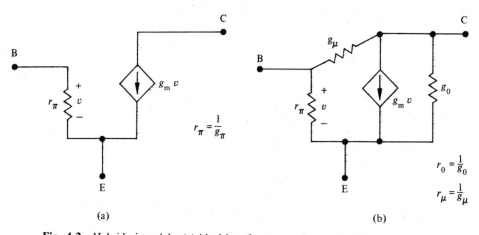

**Fig. 4.2**   Hybrid-pi models: (a) ideal low-frequency; (b) nonideal low-frequency.

## 4.3   HIGH-FREQUENCY HYBRID-PI

The high-frequency model for a bipolar transistor can be developed along the lines established for the $p$-$n$ junction since the C–B is a reverse biased diode and the E–B is a forward biased diode. We would expect the C–B junction to have a junction (depletion) capacitance that would vary with applied voltage in the same way as a reverse biased diode. Figure 4.3 illustrates the high-frequency, hybrid-pi model with a depletion capacitance $C_{jC} = C_\mu$ between the collector and intrinsic base B (see Fig. 3.7(b)). The voltage dependence is that of a reverse biased diode

$$C_\mu = \frac{C_{\mu 0}}{\left[ 1 - \dfrac{V_{CB}}{V_{bic}} \right]^m} = C_{jC} \tag{4.23}$$

where

$C_{\mu 0}$ — zero biased capacitance, similar to $C_{J0}$ for the $p^+$-$n$ junction;

$V_{bic}$ — built-in potential of the B–C junction, identical to $V_{bi}$ for the diode;

$m$ is ½ to ⅓ depending on the impurity grading of the junction.

The E–B junction has a similar depletion capacitance $C_{jE}$, with definitions similar to those of Eq. (4.23)

$$C_{jE} = \frac{C_{jE0}}{\left[ 1 - \dfrac{V_{EB}}{V_{bie}} \right]^m} \tag{4.24}$$

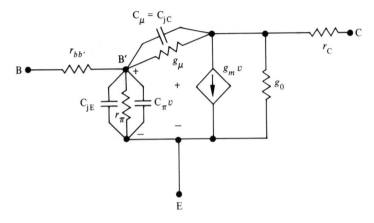

**Fig. 4.3** High-frequency hybrid-pi model.

where

$$C_{jE0} - \text{zero biased depletion capacitance;}$$

$$V_{bie} - \text{built-in potential of E–B junction;}$$

$$m \text{ is } \tfrac{1}{2} \text{ to } \tfrac{1}{3} \text{ depending on the impurity grading.}$$

The extrinsic base resistance, $r_{bb'}$, of Fig. 4.3 has been discussed previously (in Chapter 3) and is primarily the metal-semiconductor contact resistance plus any bulk base region resistance. Resistor $r_C$ in Fig. 4.3 is a similar bulk and contact resistance to the collector region. A well designed device minimizes $r_{bb'}$, and $r_C$.

The emitter base junction is a forward biased junction and will have, in addition to $C_{jE}$, a diffusion capacitance similar to that of the forward biased diode. In the derivation of the diffusion capacitance of the diode $C_D$, the $n$-region was considered to have a diffusion length much greater than a minority carrier diffusion length. For the $pnp$ transistor the base region is much less than a diffusion length and we must modify our capacitance accordingly. A $p^+$-$n$ diode at frequencies where $\omega\tau_p < 1/2$ has

$$C_D \cong \frac{G_0\tau_p}{2} = \frac{qI}{kT}\frac{\tau_p}{2}$$

The minority carrier lifetime is the average time a minority carrier remained alive before recombining. In a transistor we modify that to be the average time it takes a minority carrier to traverse the base width $W$, called the *base transit time* $\tau_t$. The E–B diffusion capacitance is then

$$C_\pi = \frac{qI_E}{kT}\tau_t \cong g_m\tau_t \tag{4.25}$$

In most cases, where the E–B is reasonably forward biased $C_\pi$ is much larger than $C_{jE}$.

The complete hybrid-pi model of Fig. 4.3 is more than adequate for most devices up to frequencies of several hundred megahertz. At larger frequencies the model does not have enough phase shift between the base and collector current. This is mainly due to our assumption of $\omega\tau < 1$. We encounter the same problem as with the diode; that is, the admittance is complex and $C_\pi$ varies with frequency. The base region should be modeled similar to a lossey transmission line and will result in an additional phase shift to $i_c$. One might attempt to model $C_\pi$ with a frequency dependent capacitor; however the circuit analysis of such a device becomes a nightmare.

## 4.4  SUMMARY

The most commonly used small signal model for a bipolar transistor is the hybrid-pi model, which relates the input signals $i_b$ and $v_{eb}$ to the output signals $i_c$ and $v_{cc}$. By assuming that the carriers behave quasi-statically, the low-frequency model can be derived with the aid of a Taylor series expansion of the Ebers–Moll equations. Active region operation models, which are the most useful, allowed us to simplify the Ebers–Moll equations and avoided generating complex mathematical equations. The ideal low-frequency model assumed no base width modulation; hence $W$ and $I_{F0}$ are constants independent of the junction voltages. The ideal model was described with only two circuit elements, a resistance $r_\pi$ and a transconductance $g_m$.

The low-frequency model with base width modulation has two additional elements, namely $g_0$ and $g_\mu$, the output conductance and feedback conductance, respectively. Both of these elements represent deviations from the ideal.

A high-frequency model was obtained by extending the low-frequency model to contain capacitances. The C–B was modeled with a depletion capacitance, the E–B with a depletion and diffusion capacitance. Two bulk and contact resistances were added in series with the base and collector to complete the model. For frequencies typically less than several hundred megahertz the hybrid-pi model is an excellent representation of the real device.

## PROBLEMS

**4.1** Given $I_{F_0} \cong qAD_B p_{B0}/W$ and $W = W_B - K[-V_{CB}]^{1/2}$, derive equations for the low-frequency hybrid-pi model circuit elements $g_m$ and $g_o$. The device is otherwise ideal.

**4.2** Given $I_E = 1\,$mA, $\beta_F = 150$, $C_{\mu0} = 2\,$pF, $C_{JEO} = 10\,$pF, $V_{bic} = 0.7$, $V_{bie} = 0.9$, $m = 1/2$, and $\tau_t = 10^{-8}\,$sec, determine the high-frequency, hybrid-pi model for $V_{EB} = 0.35$ and $V_{CB} = -3$. Given $I_E = 0.1\,$mA and $V_{CB} = -5\,$V, determine the new model.

**4.3** Consider the $p^+$-$n$ emitter base junction where $V_{EB} > 0$ and $V_{CB} = 0$.

(a) determine $Q_B$, the excess base charge;

(b) determine $I_E$ if the back injected electrons are neglected;

(c) $Q_B$ for the short base region is $Q_B = I_E \tau_t$, therefore derive an equation for $\tau_t$ as some function of $W$ and $D_B$.

# 5 / Switching Transients

Digital electronics requires that the transistor be switched rapidly from cutoff, through the active region, and into saturation. After a time lapse, the transistor is switched from saturation and returned to the cutoff region of operation. The device physics involved in such large signal transients affects the speed of switching and therefore the circuit design of bipolar logic circuits. To be more specific, the speed at which a logic element can be clocked is determined by how fast the device can be made to switch and by the length of any time delays in the signal path. The goals for this chapter are to develop a model for determining the minority carrier charge storage applicable to large signal switching and to derive the terminal currents as a function of time for the turn-on and turn-off transient.

## 5.1  CHARGE CONTROL MODEL

The derivation of the minority carrier concentrations for a *pnp* device, as presented in Chapter 2, Section 2.3, results in $p_B(x)$, as illustrated by Fig. 5.1(a) for the cutoff, saturation, and active regions. In order to simplify the derivation and concentrate on the most important concepts of the device physics of switching, we assume a $p^+np^+$ device. By using a heavily doped emitter and collector we force the base minority carrier concentration to be much larger than the minority carrier electrons in the $p^+$-bulk regions, as shown in Fig. 5.1(b) for the case of saturation. Therefore, when switching from saturation to cutoff, or vice versa, the largest charge component that must be changed is the holes stored in the base region, as the cross-hatched areas of Fig. 5.1(b) indicate. We assume the $p^+np^+$ bipolar transistor to be approximated by the hole concentration illustrated by Fig. 5.1(c). Note that because $\Delta p_B(x)$ is approximately a straight line, the excess hole concentration can be broken into two parts. The first part, labeled $Q_N$, is the total excess charge in the base when $V_{CB} = 0$ and $V_{EB} > 0$. The second is $Q_I$, the total excess charge in the base when $V_{CB} > 0$ and $V_{EB} = 0$. If the total hole charge in the base

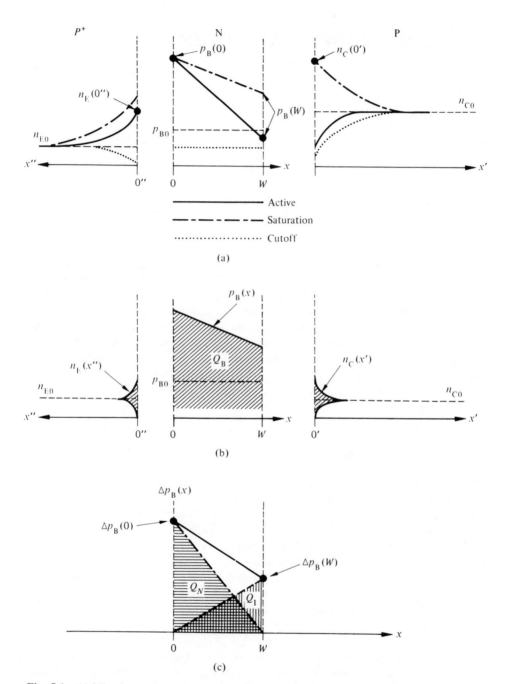

**Fig. 5.1**   (a) Minority carrier concentration $p^+np$; (b) $p^+np^+$ charge removal in switching from the saturation to the cutoff region; (c) base charge components in the saturation region.

is $Q_B$, then

$$Q_N = qA\frac{W}{2}\Delta p_B(0) \tag{5.1}$$

$$Q_I = qA\frac{W}{2}\Delta p_B(W) \tag{5.2}$$

$$Q_B = Q_N + Q_I \tag{5.3}$$

Remember that the boundary conditions at the edges of the base bulk region are $\Delta p_B(0) = p_{B0}(e^{qV_{EB}/kT} - 1)$ and $\Delta p_B(W) = p_{B0}(e^{qV_{CB}/kT} - 1)$.

For the $p^+np^+$ device in saturation, the emitter current is primarily determined by the slope of $\Delta p_B(x)$ at $x = 0$, and the collector current by the slope at $x = W$ because the electron components are very small due to the $p^+$-$n$ and $n$-$p^+$ junctions. The base currents due to the back injected electrons from base to the $p^+$ regions are also negligible. Therefore, base current is primarily that due to recombination of holes in the base; that is, $I_B \cong Q_B/\tau_B$.

A common emitter amplifier has $i_B(t)$ as the input variable that, in our case, controls the base charge storage. The rate of change of $Q_B(t)$ at any instant of time is determined by $i_B(t)$ adding charge to the base region and by recombination removing charge. Arguments similar to those for the $p^+$-$n$ diode (see Section 6.2, Vol. II) indicate the base charge to be determined by Eq. (5.4).

$$\frac{dQ_B(t)}{dt} = i_B(t) - \frac{Q_B(t)}{\tau_B} \tag{5.4}$$

The base charge is changed by diffusion current and recombination.

The collector current, $i_C(t)$, is the output variable of a common emitter connected bipolar transistor. In the case of active region operation $\Delta p_B(W) \cong 0$ and $Q_B \cong Q_N$. The collector current can be determined by the total charge in the base that must be transferred to the collector every $\tau_t$ seconds, where $\tau_t$ is defined as the *base transit time*.* The collector current is

$$i_C(t) = \frac{Q_N(t)}{\tau_t} = \frac{Q_B(t)}{\tau_t}, \qquad \text{active region} \tag{5.5}$$

Inspection of Fig. 5.1(c) for a $p^+np^+$ device at the edge of saturation, where $V_{CB} = 0$ and $V_{EB} > 0$, is the basis for defining the *base charge at the edge of saturation*, $Q_{sat}$, as

$$Q_{sat} = Q_N = I_{Csat}\tau_t, \qquad \text{edge of saturation} \tag{5.6}$$

where any $V_{CB} > 0$ requires $Q_B > Q_{sat}$. Note that a larger $I_{Csat}$ requires a larger $Q_B = Q_{sat}$.

---

*The transit time, $\tau_t$, is also the average time that it takes a carrier to transverse the base region.

## 5.2   TURN-ON TRANSIENT

The switching of a transistor from cutoff to saturation is called the *turn-on transient*. Typically this is accomplished by using a circuit similar to Fig. 5.2. With $V_S$ at zero or some negative value, the transistor is "off," or in the cutoff region of operation, and $v_{EC} \cong V_{CC}$ with $i_C \cong 0$ as indicated by point A in Fig. 5.2(b). When $V_S$ is pulsed positive, the base current increases and $v_{EC}$ decreases towards point B, from cutoff through the active region and into saturation.

Consider the case where $V_S \gg v_{EB}$; then $i_B \cong V_S/R_S = I_B$. That is, the base current is a constant. Applying Eq. (5.4) yields

$$\frac{dQ_B(t)}{dt} = I_B - \frac{Q_B(t)}{\tau_B} \tag{5.7}$$

and if $Q_B = 0$ before the base current is switched, the solution of Eq. (5.7) results in Eq. (5.8).

$$Q_B(t) = I_B \tau_B (1 - e^{-t/\tau_B}) \tag{5.8}$$

A plot of $Q_B(t)$ is shown in Fig. 5.2(c). Note that for $Q_B < Q_{sat}$, the transistor is switching through the active region on its way to saturation and $Q_B(t) \cong Q_N(t)$. The collector current, $i_C(t)$, is obtained from Eq. (5.5) as

$$i_C(t) = \frac{Q_B(t)}{\tau_t} = \frac{I_B \tau_B}{\tau_t}(1 - e^{-t/\tau_B}), \qquad Q_B \le Q_{sat} \tag{5.9}$$

For most devices $\tau_B$ is much larger than $\tau_t$. In fact, as $t \to \infty$, $i_C \to I_C$, and from Eq. (5.9), $I_C/I_B = \tau_B/\tau_t = \beta_F$, provided $Q_B(\infty)$ does not reach $Q_{sat}$ and the device remains active. The general case for $i_C$ is plotted in Fig. 5.2(c), where $t_r$ is the time it takes to reach the edge of saturation. We can solve for $t_r$ by equating $I_{Csat}$ from Eq. (5.6) to Eq. (5.9) with $t = t_r$.

$$I_{Csat} = \frac{Q_{sat}}{\tau_t} = \frac{I_B \tau_B}{\tau_t}(1 - e^{-t_r/\tau_B}) \tag{5.10}$$

$$t_r = \tau_B \ln\left[\frac{1}{1 - \dfrac{I_{Csat}\tau_t}{I_B\tau_B}}\right] \tag{5.11}$$

where

$$I_{Csat} = \frac{V_{CC} - V_{ECsat}}{R_L} \cong V_{CC}/R_L \tag{5.12}$$

and

$$I_B = \frac{V_S - v_{EC}}{R_S} \cong V_S/R_S \tag{5.13}$$

Inspection of Eq. (5.11) indicates that a smaller $I_{Csat}$ and/or a larger $I_B$ will reduce $t_r$, as will a smaller value of $\tau_B$.

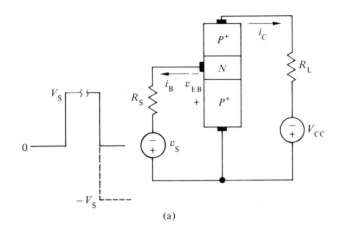

(a)

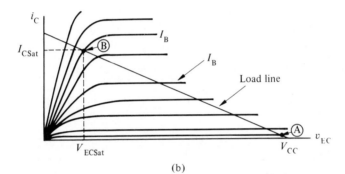

(b)

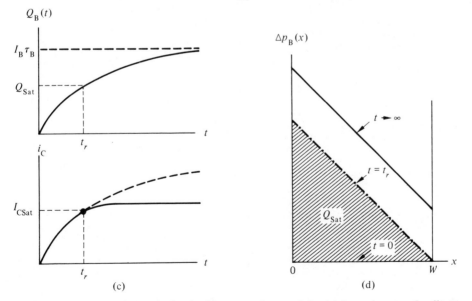

(c)                                                        (d)

**Fig. 5.2**  Turn-on transient: (a) circuit; (b) output characteristic; (c) base charge and collector current; (d) excess base charge.

Note in Fig. 5.2(c) that once $Q_B$ reaches $Q_{sat}$, the collector current increases only a small amount. One way to determine that $I_{Csat}$ is approximately constant is illustrated in Fig. 5.2(d), where the slope of $\Delta p_B(W)$ does not change significantly once in saturation. Another is Fig. 5.2(b), where once in saturation further increases in $I_B$ result in only a slight change in $i_C$ or $v_{EC}$.

## 5.3  TURN-OFF TRANSIENT

The switching of a bipolar transistor from the saturation region, through the active region, to cutoff is called the *turn-off transient*. For the common emitter of Fig. 5.1(a) the base current is switched from a value of $I_B$ (in saturation) either to zero or to some negative $i_B$.

### 5.3.1  $I_B$ to 0

The base current is switched from $I_B$, where the device is saturated, to zero as $V_S$ is switched to zero, as illustrated in Fig. 5.3(a). Applying Eq. (5.6) we obtain Eq. (5.14),

$$\frac{dQ_B(t)}{dt} = 0 - \frac{Q_B(t)}{\tau_B} \tag{5.14}$$

which has a solution

$$Q_B(t) = Q_B(0)e^{-t/\tau_B} \tag{5.15}$$

and is plotted in Fig. 5.3(b). Here $Q_B(0)$ is the total charge under the $t = 0$ line in Fig. 5.3(c), and the further the device is initially into saturation, the larger $Q_B(0)$.

The collector current remains relatively unchanged until $Q_B(t)$ is reduced to $Q_{sat}$, the time of which is defined as the *storage time delay*, $t_{sd}$. When $Q_B(t) = Q_{sat}$, the device is at the edge of the saturation active region, that is, $Q_B(t_{sd}) = Q_{sat}$. For further time increases the device is in the active region and

$$i_C(t) = \frac{Q_B(t)}{\tau_t} = \frac{Q_B(0)}{\tau_t}e^{-t/\tau_B} \tag{5.16}$$

as plotted in Fig. 5.3(d).

The storage time can be determined from Eq. (5.16) by equating $I_{Csat}$ to $i_C(t_{sd})$,

$$I_{Csat} = \frac{Q_B(0)}{\tau_t}e^{-t_{sd}/\tau_B} = \frac{I_B \tau_B}{\tau_t}e^{-t_{sd}/\tau_B} \tag{5.17}$$

and solving for $t_{sd}$ yields Eq. (5.18).

$$t_{sd} = \tau_B \ln\left[\frac{I_B \tau_B}{I_{Csat}\tau_t}\right] \tag{5.18}$$

Note that a smaller $\tau_B$ or $I_B$ will reduce $t_{sd}$; that is, the less the device is pushed into saturation and/or the smaller the stored base charge, the shorter the storage time delay. Once in the active region, $i_C(t)$ decays exponentially toward zero as shown in Fig. 5.3(d).

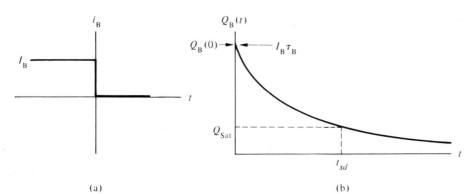

(a)

(b)

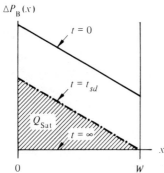

(c)

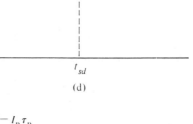

(d)

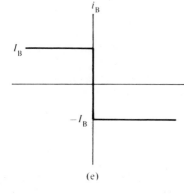

(e)

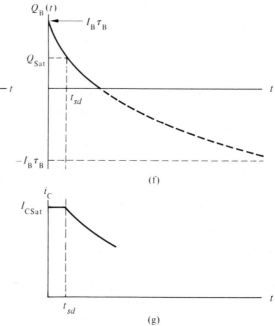

(f)

(g)

**Fig. 5.3**   (a) $i_B = 0$ for $t \geq 0$; (b) $Q_B(t)$ for $i_B = 0$; (c) $\Delta p_B(x)$; (d) $i_C(t)$ for $i_B = 0$; (e) $i_B = -I_B$ for $t \geq 0$; (f) $Q_B(t)$ for $i_B = -I_B$; (g) $i_C(t) = -I_B$.

Because of $t_{sd} = 0$ for nonsaturating logic circuits, they are switched faster than saturating circuits.

## 5.3.2  $I_B$ to $-I_B$

To speed up the removal of minority carrier charge from the base and hence reduce the switching time, the base current is reversed to a value of $-I_B$, as illustrated in Fig. 5.3(e). Equation (5.4) becomes

$$\frac{dQ_B(t)}{dt} = -I_B - \frac{Q_B}{\tau_B} \tag{5.19}$$

which has a solution Eq. (5.20).

$$Q_B(t) = I_B \tau_B (2e^{-t/\tau_B} - 1) \tag{5.20}$$

Equation (5.20) is plotted in Fig. 5.3(f). While in saturation, $Q_B(t) > Q_{sat}$, the collector current is nearly constant, as illustrated in Fig. 5.3(g). When $Q_B(t) = Q_{sat}$, then $i_C(t) = I_{Csat}$ and $t = t_{sd}$.

$$I_{Csat} = \frac{Q_{sat}}{\tau_t} = \frac{I_B \tau_B}{\tau_t} (2e^{-t_{sd}/\tau_B} - 1) \tag{5.21}$$

With some algebraic manipulation the storage time delay is

$$t_{sd} = \tau_B \ln \left\{ \frac{I_B \tau_B}{I_{Csat} \tau_t \left[ 1/2 + 1/2 \dfrac{I_B \tau_B}{I_{Csat} \tau_t} \right]} \right\} \tag{5.22}$$

Comparing Eq. (5.22) with Eq. (5.18) we find that $t_{sd}$ has been reduced by using a negative base current $(-I_B)$ to aid in removing $Q_B$ from the base region. In Fig. 5.3(f), $Q_B(t)$ is attempting to reach a value of $-I_B \tau_B$; however, our solution ends when $Q_B \cong 0$. Also note that in the active region the collector current is decreasing at a faster rate than in the previous case, when discharging into $I_B = 0$.

## 5.4  SUMMARY

The charge control model for the switching of a $p^+np^+$ device was derived in terms of the base region, minority carrier charge $Q_B(t)$. By assuming $p^+$ emitter and collector regions the minority carrier charge storage in their bulk regions is insignificant compared to the base bulk region and the base current can be approximated as the recombination component, $Q_B/\tau_B$. The base charge storage is controlled by the base current adding holes and by recombination removing holes.

   A turn-on transient response was modeled, for the common emitter, from cutoff into saturation by solving for the hole charge in the base $Q_B(t)$. At the edge of saturation, $Q_B = Q_{sat}$ and $i_C = I_{Csat}$. Once into saturation, $i_C$ is relatively constant. The turn-on time can be reduced by increasing the base current drive $I_B$ and reducing the base lifetime $\tau_B$.

The turn-off transient is characterized by two time intervals. One is the storage time, $t_{sd}$, which is the time required to remove a sufficient amount of base charge to bring the device from deep saturation to the edge of the active region. Once in the active region the collector current can respond to the signal and begins to decrease, which is the second time interval. When switching the transistor with a negative base current, $-I_B$, the charge removal is faster than that when switching with a zero base current; hence there is a shorter switching time. In summary, the turn-off transient is reduced by having less $Q_B$ to change. Therefore the less the device is pushed into saturation, the shorter the storage time.

## PROBLEMS

**5.1** If $\tau_B = 0.1$ $\mu$sec and $\tau_t = 1$ n sec, plot $t_r/\tau_B$ versus ln $I_B$ for the case of $I_{Csat} = 10$ mA. Remember $t_r$ is the time at which the device is saturated. *Hint:* let 100 $\mu$A $\leq I_B \leq$ 100 mA. What do you conclude about the base drive?

**5.2** If $\tau_B = 0.1$ $\mu$sec and $\tau_t = 1$ n sec, plot $t_{sd}/\tau_B$ versus ln $I_B$ for $I_{Csat} = 10$ mA. Let 100 $\mu$A $\leq I_B \leq$ 100 mA.

(a) for the case of switching $i_B$ to 0;

(b) for the case of switching $i_B$ to $-I_B$.

**5.3** Derive the storage time delay if $i_B$ is switched from $I_{B_1}$ to $-I_{B_2}$.

# Suggested Readings

1. A. B. Glaser and G. E. Subak-Sharpe, *Integrated Circuit Engineering*. Reading, Mass.: Addison-Wesley, 1977. Chapter 2 presents a detailed picture of the junction capacitance, breakdown voltage, and integrated circuit modeling of the bipolar transistor. The base spreading resistance and three-dimensional effects are discussed in detail.

2. W. H. Hayt and G. W. Neudeck, *Electronic Circuit Analysis and Design*. Boston: Houghton Mifflin Co., 1976. Chapter 2 is an elementary physical description of the device and its important volt–ampere characteristics. Chapters 3 through 9 are circuit applications.

3. J. L. Moll, *Physics of Semiconductors*. New York: McGraw–Hill, 1964. Chapter 8 contains a formal discussion of the high-frequency beta and charge storage phenomena.

4. B. G. Streetman, *Solid State Electronic Devices*. 2nd ed., Englewood Cliffs, N. J.: Prentice–Hall, 1980. Chapter 7 is a thorough discussion of the bipolar transistor including the deviations from ideal.

# Volume Review Problems and Answers

## VOLUME REVIEW PROBLEMS

**A.1** List the regions of operation for a bipolar transistor.

**A.2** For a *pnp* device indicate the voltage polarity (+ or −) for the following:

| Region | $V_{EB}$ | $V_{CB}$ |
|---|---|---|
| Active | | |
| Saturation | | |
| Cutoff | | |
| Inverted active | | |
| Inverted saturation | | |

**A.3** Sketch the thermal equilibrium energy band diagram for a *pnp* transistor.

**A.4** Sketch the current components and indicate the electron and hole current flow for an active region (a) *pnp* (b) *npn*.

**A.5** Explain "diode isolation."

**A.6** Define the base transport factor for an *npn* device in terms of $I_{Cn}$ and $I_{En}$. What prevents this from being unity?

**A.7** Given that the ideal *pnp* has an emitter injection efficiency of 0.99 and the C–B leakage current is 10 microamperes, calculate the active region emitter current due to holes if $I_B = 0$.

**A.8** Sketch the minority carrier distributions for saturation-region operation of a nonideal $p^+np$.

**A.9** For a quasi-ideal *pnp*, sketch the minority carrier distributions for inverted operation.

**A.10** Explain the origin of the three base current components for a *pnp* device.

**A.11** How are the Ebers–Moll equations modified for an *npn* device? Write the equations and draw an equivalent circuit.

**A.12** Why is the emitter doped more heavily than the base?

**A.13** Are $\alpha_{dc}$ and gamma identical for a quasi-ideal device? Explain.

**A.14** Use a carrier flow diagram to explain why $I_{CEO}$ is much different from $I_{CBO}$ for an *npn*.

**A.15** The "base region is small" is often stated in device analysis. What is "small" and what is it compared to?

**A.16** Sketch the particle fluxes and their relative sizes for an ideal $p^+np$ device operating in (a) saturation, (b) cutoff, (c) inverted active.

**A.17** If $W$ were made smaller, explain how it would affect the base width modulation.

**A.18** Why does $\beta_{dc}$ increase with increasing $I_C$ at small values of collector current?

**A.19** Explain why beta falls off at large values of collector current.

**A.20** Sketch the low-frequency, nonideal, hybrid-pi model. How will the transconductance vary with $T$?

**A.21** Sketch $C_\mu$ versus $V_{CB}$ and $C_{jE}$ versus $V_{BE}$.

**A.22** If base width modulation effects were made greater, how would this affect the values of the hybrid-pi model elements?

**A.23** Sketch the minority carrier charge storage changes required for a *pnp* device to be switched from cutoff to saturation. Use a cross-hatched area to represent the total change.

**A.24** If the base charge is quadrupled, what is the difference between the two storage-time delays when the base current is switched to zero?

**A.25** List three methods used to reduce the turn-off time.

## ANSWERS TO VOLUME REVIEW PROBLEMS

**A.1**   Active, saturation, cutoff, inverted

**A.2**

| Region | $V_{EB}$ | $V_{CB}$ |
|---|---|---|
| Active | + | − |
| Saturation | + | + |
| Cutoff | − | − |
| Inverted active | − | + |
| Inverted saturation | + | + |

**A.3**

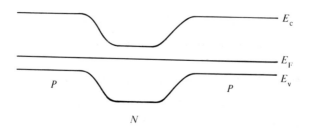

**Figure A.3**

**A.4**

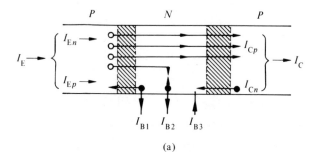

(a)

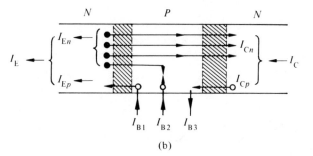

(b)

**Figure A.4**

**A.5**    diode isolation.

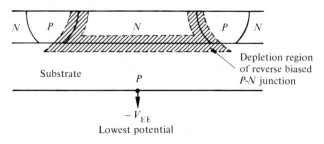

**Figure A.5**

**A.6**    $\alpha_T = I_{Cn}/I_{En}$; any electrons that are recombined in the base.

**A.7**    $\beta_{dc} = 99$ and $I_{CE0} = 100$ nA

**A.8**

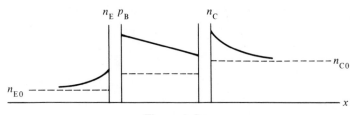

**Figure A.8**

**A.9**

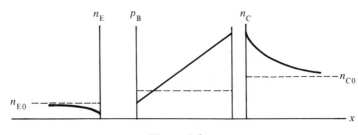

**Figure A.9**

**A.10** $I_{B_1}$ is the back injected electrons from the B–E. $I_{B_2}$ is the current supplying electrons for recombination with holes in $W$. $I_{B_3}$ is the electron current from C–B falling down the potential hill; it is thermally generated within one $L_C$ of the depletion region edge.

**A.11** $p$ becomes $n$ and $n$ becomes $p$; $V_{BE}$ becomes $V_{EB}$; all $I$'s change direction.

$$I_E = I_F - \alpha_R I_R$$
$$I_C = \alpha_F I_F - I_R$$
$$I_B = (1 - \alpha_F) I_F + (1 - \alpha_R) I_R$$

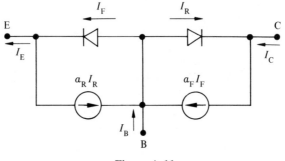

**Figure A.11**

**A.12** To improve the emitter injection efficiency; reduce the back injected carriers from B–E.

**A.13** No, because some of the holes are lost to recombination in $W$; hence $I_{Cp}$ is unequal to $I_{Ep}$ and $\alpha_T < 1$. However, since the recombination current is small compared to $I_C$ and $I_E$, it is a good approximation that $\alpha_{dc}$ is almost $\gamma$.

**A.14** $I_{CBO} \approx I_{Cp}$, provide back injected holes for E–B, causes E–B to be forward biased.

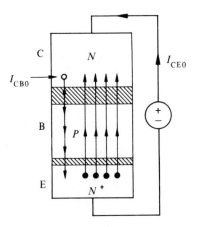

**Figure A.14**

**A.15**   Small means $W \ll L_B$, typically $\leq \mu$

**A.16**

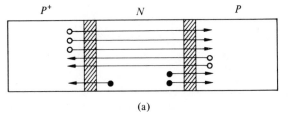

(a)

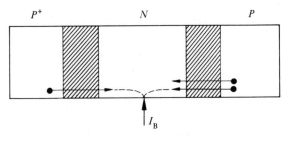

(b)

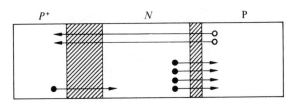

(c)

**A.17** $\Delta W$ becomes a larger percentage of $W$; therefore it increases the slope of output characteristics.

**A.18** Due to recombination current in the E–B depletion region, which becomes a smaller part of $I_B$ as $I_E$ gets larger.

**A.19** At larger $I_C$ high injection (or $\tau_C$) not to increase exponentially.

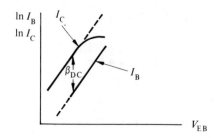

**Figure A.19**

**A.20** $g_\pi = q(I_C/kT)$; therefore $T^{-1}$ if $I_C$ is constant.

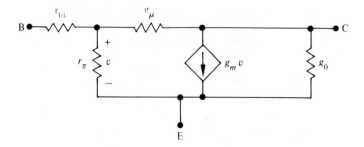

**Figure A.20**

**A.21**

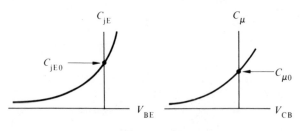

**Figure A.21**

**A.22** $g_\mu$ and $g_0$ get larger.

**A.23**

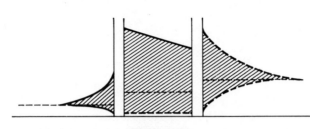

**Figure A.23**

**A.24** $\tau_B \ln 4$

**A.25** $\tau_B$, $I_B$, $I_{Csat}$ decrease, $I_B$ larger.

# Appendix

## LIST OF SYMBOLS

| | |
|---|---|
| $BV_{CB0}$ | breakdown voltage of C–B with $I_E = 0$ |
| $BV_{CE0}$ | breakdown voltage of C–E with $I_B = 0$ |
| $BV_{CER}$ | breakdown voltage of C–E with $R$ from B–E |
| $C_\mu$ | junction (depletion) capacitance of C–B junction |
| $C_{\mu 0}$ | junction (depletion) capacitance of C–B junction $V_{BC} = 0$ |
| $C_{jC}$ | junction (depletion) capacitance of C–B junction |
| $C_{jE}$ | junction (depletion) capacitance of E–B junction |
| $C_{JE0}$ | junction (depletion) capacitance of E–B junction $V_{EB} = 0$ |
| $C_D$ | diode diffusion capacitance |
| $C_\pi$ | E–B diffusion capacitance |
| $D_N$ | electron diffusion constant |
| $D_P$ | hole diffusion constant |
| $D_E$ | emitter minority carrier diffusion constant |
| $D_B$ | base minority carrier diffusion constant |
| $D_C$ | collector minority carrier diffusion constant |
| $E_C$ | conduction band edge energy level (eV) |
| $E_i$ | intrinsic energy level (eV) |
| $E_F$ | Fermi energy level (eV) |
| $E_V$ | valence band edge energy level (eV) |
| $g_m$ | low-frequency signal transconductance |
| $g_0$ | low-frequency signal output conductance |
| $g_\pi$ | low-frequency signal input conductance |
| $g_\mu$ | low-frequency signal feedback conductance |

| | |
|---|---|
| $G_D$ | low-frequency diode signal conductance |
| $I_B$ | dc base current |
| $I_C$ | dc collector current |
| $I_E$ | dc emitter current |
| $i_B$ | total instantaneous base current |
| $i_C$ | total instantaneous collector current |
| $i_E$ | total instantaneous emitter current |
| $i_b$ | signal base current |
| $i_c$ | signal collector current |
| $i_e$ | signal emitter current |
| $I_N$ | electron current |
| $I_P$ | hole current |
| $I_{B_1}$ | base current component due to back-injected carriers |
| $I_{B_2}$ | base current component due to recombination |
| $I_{B_3}$ | base current component due to carriers from the collector |
| $I_{C_{sat}}$ | collector current at edge of saturation |
| $I_{Ep}$ | emitter current due to holes |
| $I_{En}$ | emitter current due to electrons |
| $I_{Cp}$ | collector current due to holes |
| $I_{Cn}$ | collector current due to electrons |
| $I_{CB0}$ or $I_{BC0}$ | C–B leakage current, $I_B = 0$ |
| $I_{CE0}$ or $I_{EC0}$ | C–E leakage current, $I_B = 0$ |
| $I_F$ | Ebers–Moll forward current component |
| $I_{F0}$ | Ebers–Moll forward coefficient |
| $I_R$ | Ebers–Moll reverse current component |
| $I_{R0}$ | Ebers–Moll reverse coefficient |
| $I_S = \alpha_F I_{F0} = \alpha_R I_{R0}$ | |
| $L_N$ | electron diffusion length (cm) |
| $L_P$ | hole diffusion length (cm) |
| $L_E$ | emitter minority carrier diffusion length |
| $L_B$ | base minority carrier diffusion length |
| $L_C$ | collector minority carrier diffusion length |
| $m$ | coefficient of depletion capacitance, $\frac{1}{3} \leqq m \leqq \frac{1}{2}$ |
| $N_A$ | acceptor impurity concentration ($\#/cm^3$) |
| $n^+$ | degenerately doped $n$-type material |
| $N_D$ | donor impurity concentration ($\#/cm^3$) |
| $N_{AE}$ | acceptor concentration of the emitter |

| | |
|---|---|
| $N_{DB}$ | donor concentration of the base |
| $N_{AC}$ | acceptor concentration of the collector |
| $n$ | ideality factor of junction |
| $n_p$ | electron concentration in $p$-material |
| $n_{p0}$ | electron concentration in $p$-material at thermal equilibrium |
| $n_{n0}$ | electron concentration in $n$-material at thermal equilibrium |
| $n_n$ | electron concentration in $n$-material |
| $n_{E0}$ | electron concentration in emitter at thermal equilibrium |
| $n_{C0}$ | electron concentration in collector at thermal equilibrium |
| $n_E$ | electron concentration in emitter |
| $n_C$ | electron concentration in collector |
| $p^+$ | degenerately doped $p$-type material |
| $p_p$ | hole concentration in $p$-type material |
| $p_{p0}$ | hole concentration in $p$-type material at thermal equilibrium |
| $p_n$ | hole concentration in $n$-type material |
| $p_{n0}$ | hole concentration in $n$-type material at thermal equilibrium |
| $p_{B0}$ | hole concentration in base at thermal equilibrium |
| $p_B(x)$ | hole concentration in base |
| $Q_B$ | minority carrier charge storage in base |
| $Q_N$ | normal minority carrier charge storage in base, $V_{CB} = 0$, $V_{EB} > 0$ |
| $Q_I$ | inverted minority carrier charge storage in base, $V_{EB} = 0$, $V_{CB} > 0$ |
| $Q_{sat}$ | base minority charge storage at edge of saturation |
| $r_0 = 1/g_0$ | low-frequency signal output resistance |
| $r_{bb'}$ | extrinsic base resistance |
| $r_C$ | collector bulk and contact resistance |
| $t$ | time |
| $t_r$ | rise time of collector current |
| $t_{sd}$ | storage time delay |
| $V_{EB}$ | emitter-to-base voltage |
| $V_{CB}$ | collector-to-base voltage |
| $V_{EC}$ | emitter-to-collector voltage |
| $v_{EB}$ | total instantaneous E–B voltage |
| $v_{CB}$ | total instantaneous C–B voltage |
| $v_{EC}$ | total instantaneous E–C voltage |
| $v_{eb}$ | signal E–B voltage |
| $v_{cb}$ | signal C–B voltage |
| $v_{ec}$ | signal E–C voltage |

| | |
|---|---|
| $V_{bic}$ | built-in potential of B–C junction |
| $V_{bie}$ | built-in potential of E–B junction |
| $W$ | base width (bulk region) (cm) |
| $W_B$ | metallurgical base width |
| $x, x', x''$ | $x$-axis variables |
| $\rho$ | charge density (coul/cm$^3$) |
| $\mathscr{E}$ | electric field (V/cm) |
| $\alpha_{dc}$ | dc alpha |
| $\alpha_T$ | base transport factor |
| $\gamma$ | emitter injection efficiency |
| $\beta_{dc}$ | dc beta |
| $\beta_F$ | normal forward beta, Ebers–Moll |
| $\beta_R$ | reverse beta, Ebers–Moll |
| $\Delta V_{EB}$ | signal component of E–B voltage |
| $\Delta n_E$ | excess electron concentration in emitter |
| $\Delta n_C$ | excess electron concentration in collector |
| $\Delta p_B$ | excess hole concentration in base |
| $\tau_E$ | minority carrier lifetime in emitter |
| $\tau_B$ | minority carrier lifetime in base |
| $\tau_C$ | minority carrier lifetime in collector |
| $\alpha_F$ | forward alpha, Ebers–Moll |
| $\alpha_R$ | reverse alpha, alpha, Ebers–Moll |
| $\tau_p$ | hole minority carrier lifetime |
| $\tau_t$ | base transit time |
| $\omega$ | radian frequency |

# Index